Fiona Stafford is Professor of English at the University of Oxford. She specialises in literature of the Romantic period (especially Wordsworth, Austen, Burns, Keats, Clare), Scottish and Irish literature, contemporary poetry, environmental humanities and nature writing, literature and the visual arts. In addition to academic books and essays, she contributes to newspapers, literary magazines, art books, Radio 3's *The Essay* and collections of nature writing. She is the author of *The Long, Long Life of Trees*, and *Jane Austen: A Brief Life*.

Also by Fiona Stafford

The Long, Long Life of Trees

The Brief Life of Flowers

FIONA STAFFORD

JOHN MURRAY

First published in Great Britain in 2018 by John Murray (Publishers)
An Hachette UK company

This paperback edition published in 2019

4

A CIP catalogue record for this title is
available from the British Library

ISBN 978-1-47368-637-3
Ebook ISBN 978-1-47368-636-6

Typeset in Bembo MT by Palimpsest Book Production Limited,
Falkirk, Stirlingshire

Printed and bound in Great Britain by Clays Ltd, Elcograf S.p.A.

John Murray policy is to use papers that are natural, renewable and
recyclable products and made from wood grown in sustainable forests.
The logging and manufacturing processes are expected to conform
to the environmental regulations of the country of origin.

John Murray (Publishers)
Carmelite House
50 Victoria Embankment
London EC4Y 0DZ

www.johnmurray.co.uk

For my mother, Gill Stafford,
and my sister, Sue Downie

Contents

Springs

I can measure my entire life in leaves and petals. My window opens on to honeysuckles and roses, cascading over walls, clambering into trees, carefully planted to conceal the fact that only a few years ago this patch of land was a field full of broken bricks and rusty scraps of abandoned farm machinery, where the nettles and thistles felt entirely at home. The thistles and nettles still make an insistent appearance each spring, but not in such great numbers, as the space slowly fills with flowers and shrubs and young trees. The thistles form a living link between the maturing garden and the hay field over the fence, between the immediacy of this budding summer and all the summers that have gone before.

In the backyard of our Victorian brick terraced house, where my family lived sporadically during my childhood in between postings to various RAF stations, was a small, almost rectangular patch of red- and yellow-headed snapdragons, whose mouths gaped obediently when squeezed. It is hidden in my mind, somewhere behind the rose-covered, limewashed cottage in the woods where we lived during my teenage years. Once home to grooms and servants, the white cottage formed the corner of a square stable-yard, since softened with mounds of mauve aubretia and bursts of ox-eye daisies, lime-green sprays of lady's mantle, bright gladioli spikes, a great rambling clematis and a pool of cool irises and water lilies. When I think of these yards and their flowers, they are always in summer colour, because that was when we spent most time out of doors.

We moved often and so the new garden was always somewhere to explore. One early home in Yorkshire opened on to an expanse of lawn big enough to floor a four-year-old; it was hemmed in by crowds of puffy pink, white and purple lupins, and tufty, strong-smelling lavender, where enormous bees balanced impossibly. On the warm, tiled steps by the front porch, tiger-striped caterpillars wriggled their way from beneath flat, furry geranium leaves, before disappearing into low-lying, sticky clumps of something I've never been able to identify. This garden was large enough to allow each of us a little strip of earth, which we sprinkled with black powders, pale pips of nasturtium and marigold seeds curled like dried maggots. When we said goodbye to this home, we sailed across the North Sea to a new life in the Netherlands and drove for hours through vast tulip fields in primary colours, laid out like huge flags for high-flying pilots and swallows. Our Dutch house, black and white like the dairy cows, was half hidden by apple blossom and flanked by fields of buttercups. In between the waving lines of cow parsley along the verges of the narrow local roads were small white shrines, framed by fresh floral decorations.

No matter where we landed, the garden, whether long settled or recently planted, was new to our family. A weedy bed or a rubble heap was an open challenge to begin the metamorphosis. Even when my father was stationed in Aden on the Red Sea, where the dry air, volcanic rock and scorching sands meant market stalls were spread with bags of coffee, nuts and spices rather than cut flowers or bedding plants, my mother still scraped together a patch of green, nurturing a weed known as 'shameless', which seemed to grow in a spirit of unlikelihood. My grand-mother, still settled amid smooth lawns, abundant troughs of seasonal colour and sloping herbaceous borders, rapidly dispatched a box of beautiful wax flowers which, being ill-suited to survival in Yemen, began to melt into surreal sculptures and were sprayed with gritty, dull gold sand within minutes of arrival. Wherever we were, flowers were soon there too. Before the crates were unpacked and familiar old vases emerged, a jar or an unemployed jug would take on the temporary job of flower-bearer. If the new back door opened on to nothing but concrete slabs or scrubby beds, a bunch of flowers from the nearest filling station would be carried over the threshold with the first bags of essential supplies. Since my mother had trained as a florist at the Constance Spry School before her marriage, throughout the year, windowsills, bookcases and mantel-pieces would be adorned with ever-changing arrangements, from fans of daffodils to spikier, red-berried evergreen displays.

Beyond the boundaries of home we gradually dis-covered the wild flowers that marked out the surrounding terrain. In the Netherlands, the way down to the river and the fat, corrugated willows was so thick with tall nettles that, by July, it was almost impossible to reach the bank where marsh marigolds and cresses spread into the slow-moving water. When the Lincolnshire woods were bare, certain flowers were quick to get off the ground: the ice-white tips of snowdrops, the yellow celandines and the aconites appearing here and there among the ivy and dead leaves, the pale lemon primroses opening in rings around clearings where a white cherry was suddenly centre stage.

As the grass in any of the fields I've ever known starts to grow, the colours change and change again. When the hawthorn thickens into whiteness, the bright yellow shred-heads of dandelions begin to spurt from jagged green leaves below, before turning into cloudy bubbles, a shower of miniature moons settling unexpectedly, before dissolving into drenched blades of grass. Buttercups scatter specks of a slightly softer gold over the shattered globes of purple or white clover, while along the next field yellow ragwort and scarlet poppies fleck the mint-green waves of winter barley. On any day from March to September, a sudden torrent of rain can rinse or wreck fresh petals.

During years of successive student rooms, bedsits and flat-sharing, I was unable to lavish care on anything larger than spider plants, which arched from chipped mugs and sent out cream shoots, weighted with miniature replicas of themselves and dangles of hopeful roots. The diminutive size of the first garden I actually owned (with my partner and a mortgage company) was therefore quite disproportionate to its importance in the floral calendar of time and being. This tiny cottage in an Oxfordshire village had for many years been home to a woman who was evidently very fond of roses. The handkerchief patch bristled with tightly packed stems and, inside, every wall was coated in layers and layers of pink floral paper. The unmatching, attached cottage next door belonged to a retired midwife, who was kind enough to offer welcome advice on shoestring gardening, if not on interior decor. She explained how her neighbour had been in the habit of snipping short lengths from favourite plants just below a leaf bud, sticking them into the soil and gaining great satisfaction as they duly turned into viable roses. As the years passed, the small bed had become a thicket of thorns and leaves and superlatively scented blooms. As the only other plants were lilies of the valley, which packed the ground around the stems, weeding was minimal and perfume maximal, though the outlook through the winter months was rather brown.

My favourite flowers in the village were the wild primroses that grew among the gravestones, making the grass beyond the yew trees glow each spring. The poet John Drinkwater, who is buried there, recalled this apparently unremarkable spot very fondly in his autobiography, commenting that the less a place 'may seem to assert itself, the more profoundly will it possess us, instruct us, become memorable'. For him, the Piddington primroses were a match for 'the Alps or the Golden Gate', and he knew that their deep roots grew perennially from an understated always-thereness.

Flowers have a way of delivering surprises, even to those who have seen them appear in exactly the same place year after year. Their annual trick of looking new is easily accomplished – because they are new. The plants remain in place, but their flowers bud, break open, bloom broadly, and then drop away into the earth. Some plants die altogether, relying for survival on their collective ability to scatter seed, others disappear for months at a time, to pop up again from bulbs or roots hidden beneath the surface of the earth. But sometimes even these fail to reappear, whether because the underground stores have been raided, poisoned, disturbed or ploughed up, or their habitual passageways have been scorched or sealed. Perhaps the insects with which they coexist have died out, as the world above ground changes: too warm, too wet, too dry, too polluted. The fragility of flowers is evident enough in their transparent petals, delicate tendrils and gold-dusty pollen: it seems remarkable that so many do hang on to existence year after year.

The riskiness and steadfastness of flowers has helped to secure their essential place in human culture. It is not just that we need flowers to turn into fruit and vegetables, to increase and to multiply and provide food for cattle and sheep as well. Nor is it that they make the prettiest decorations, immeasurably enhancing our inner and outer lives. They are all of these things – and more. Flowers are always there at the critical moments of life: as gifts to celebrate a birth or anniversary, as bouquets to adorn a bride, as wreaths to accompany the deceased to the grave and as memorials to comfort those who mourn. Flowers are summoned to create beauty equal to the magnitude of the occasion, to recall the shared course of nature and to vanish as the momentous event begins to settle into memories and files of photos. For first-time mothers, staggered, elated, exhausted, flowers help to naturalise a world altered almost beyond recognition. So many arrived after the birth of our first child that we were almost as desperate for vases as sleep. These were kind gifts from friends and family, intended to express the deep emotions springing in response to new life. They were quietly commemorative, too, of what leads to such celebrations – those compelling feelings so often shared less publicly through flowers. A single rose can say more than a multitude of words, as poets acknowledge when resorting to the most well-worn image of all. Wedding bouquets subsume the red roses and clutched bunches of courtship, promising an eternal bower of bliss. Matrimony generally means masses of flowers, framing the entrances, adorning aisles and altars with delicate splendour, infusing the moment with natural incense. No matter how fraught the build-up to the great event, a day filled with flowers will allay all stress and strain. The paper confetti tipped over the happy couple, the posy thrown into the crowd of well-wishers, the bright garlands and scented oils adorning a Hindu bride and groom, all connect the contemporary moment to age-old, world-wide rituals of blessing and hope.

Unlike the priest, the registrar or the guests, flowers are silent witnesses, somehow ratifying and sanctifying life-altering – or life-ending – events. An unexpected death is often marked by pavements buried in bouquets, and some days later, with forlorn cellophane shrouds. In 1997, Kensington was stilled by an unfamiliar fragrance-filled atmosphere, overwhelming the stale air and exhaust fumes and gradually acquiring a faintly rotten flavour. The death of Diana, Princess of Wales, has been identified as a moment when the British people abandoned their stiff upper lips, often by those uneasy about the display. For many, laying flowers was a way of channelling grief, of expressing unarticulated feelings and reassuring each other that no one was facing bereavement alone. If the scale of the tribute was unprecedented in modern London, it was only a magnified version of the quiet, profoundly felt ceremonies that are taking place every day. The regular notices that simply state 'No Flowers' are a testament to the universal urge to respond in exactly the contrary way. Redirecting donations to good causes is rational and beneficial, but fails to accommodate one of the most deep-seated needs of the mourners. As people have always known instinctively, leaves and petals order us.

The ancient impulse is as evident in modern funerals, whether small family ceremonies or huge public events, as in the long tradition of elegies that have exhorted mourners to 'strew the laureate hearse' with 'bells and flow'rets of a thousand hues'. Percy Bysshe Shelley, rarely lost for words, still turned to pansies and 'violets, white and pied, and blue' when composing an elegy for his friend and fellow poet, John Keats, who died at the age of twenty-five. When Michael Longley commemorated a Belfast ice-cream seller, murdered in the Troubles, he gathered 'all the wild flowers of the Burren' into his poem, a stirring of twenty-one botanicals. Confronting death with flowers is a universal impulse, expressing love and respect as well as grief and, in the midst of chaotic emotion, making a sign of faith in continuing life. Flowers are often understood as emblems of the brevity of life, but they are irrepressible reminders of natural regeneration and fresh growth. Marc Riboud's iconic image from 1967, often known as 'The Ultimate Confrontation', of a young girl in a summer dress holding out a chrysanthemum to a line of bayonets, speaks of forces larger and ultimately more powerful than human destructiveness.

Flower Power, 1967. A plea for the end of the Vietnam War: peaceful protest at the Pentagon.

Flower mythology is in a state of perpetual metamorphosis. The Roman poet Ovid evidently understood the endlessly renewing nature of plants and delighted in turning beautiful youths and nymphs into flowers, which would continue to flourish in different stories, poems, paintings and films for the next two thousand years. Every collection of poetry is a gathering of different blooms: the word *anthology* is derived from ἄνθος (*Anthos*), the Greek word for a flower, while *posy* was originally synonymous with 'poesy', which meant a short motto in rhyme. Yet even many of the poems in famous anthologies, such as Archibald Wavell's *Other Men's Flowers* or Baudelaire's *Fleurs du Mal*, omit the merest mention of a bud or bloom. There are plenty that do, of course, since flowers are among the most enduring images of literature, often standing for the deepest feelings. When literal communication fails, poets will opt for the ancient language of the earth. While devotional poets of the Middle Ages tended to dwell on lilies and roses, the Cavaliers and Metaphysicals gathered them up as quickly as they could. Wordsworth listened to the tale repeated by the pansy at his feet; Keats, unable to see in the growing darkness, still conjured the full experience through the mingling scents of the fast fading violets and the coming musk rose. If Gerard Manley Hopkins celebrated bluebells, blossoms and weeds, 'long and lovely and lush', for Mary Oliver it was peonies that spoke to her of 'dampness and recklessness'.

The perpetual seasonal cycle of the world is marked by flowers. Botticelli's huge painting *Primavera*, in which the goddess Flora, robed in petal-patterned silk, stands between Chloris, whose open mouth spews flowers under the amorous pressure of the West Wind, and the three Graces, barefoot on the flower-spangled grass, is an Italian Renaissance expression of the same celebration of life and fertility that inspired the Roman feast of Flora or the Hindu celebration of Holi. David Hockney's *The Arrival of Spring in Woldgate, East Yorkshire in 2011 (twenty eleven)* replaces Botticelli's graceful allegorical figures and central red-robed goddess with a bold, cherry-coloured path through rich undergrowth bursting with golden, orange, pink and lemon flowers, slim tree trunks, blue fronds and a shower of flying green leaves to demonstrate Yorkshire's equal claim to love and vernal loveliness. If the specific date points to a personal experience of a particular spring, it also reminds us that every year offers a glorious repetition of the first spring, the *primavera*.

This is a spring that is still open to anyone – as long as modern cities maintain green spaces as they spread and spread. A line of trees along a city street, a bright border in a local park, a geometric bed in a shopping centre, a tiny roof garden and a well-watered window-box all burst into spring colour, just as surely as the poppies on a building site, the straggling buddleia beside the track.

Individual horticulturalists sometimes become stars of the screen or the Chelsea Flower Show, but most gardeners, growers and groundsmen remain as invisible to their bene-ficiaries as the florists who rise early and work feverishly to achieve an impression of natural abundance. This book is a tribute to the generations of men and women who have devoted their lives to flowers: the planters and breeders, collectors and designers, florists and foresters, artists and writers who have enhanced the lives of strangers. It is an expression of gratitude to all those who have shown their families and friends that flowers matter. It is written, most of all, in celebration of flowers, wild and cultivated, whose delicate forms are still powerful enough to stop us in our tracks and make us wonder.

The snowdrop blooms in The Temple of Flora, *a Regency celebration of the great Linnaeus and the world's most exotic plants.*

Snowdrops

They seemed to have come from nowhere, the little clutch of pale shapes, half hidden beneath the tangle of tired brown stems and flat, damp last-year's leaves. A white so bright that porcelain would seem dull by their side. I like to imagine that one of those westward-slipping, clapping, whistling lapwings that whisk up the greyest skies had broken briefly from its whirling companions to drop off a secret new year gift. But it is not the season for nesting and, in any case, these six white eggs are worryingly small, oddly elongated and rather perilously suspended. Had I been keeping a closer watch, I would have seen this quiet patch pierced by tips of green, spreading and whitening as the mornings passed. Today's thin cluster will be taller and fatter tomorrow, cracking open within days into perfectly balanced broken shells. The year is hatched in the unlikely undergrowth of January, despite grey skies, despite the puddles, mud and sodden fields, despite hard frosts and harder ponds, despite the snow, despite the falling snow. In winter woods, where the light pours through uncovered twigs and branches, the leaf litter below turns white with ice and snowdrops. In brown gardens, in empty parks, these flowers form drifts to rival whatever the clouds disgorge. It's almost as if a cloud of perfectly white smoke has settled over an unsuspecting corner. There is an old tradition that tells how Eve, heartbroken after her exile from the garden of abundance, was comforted by snowdrops – and hence one of the local names for snowdrops in Somerset is 'Eve's tears'. Perhaps the streaks of pale green on the inner whorl of white petals reminded her of the Eden she had lost; perhaps they carried the tiny promise of a fresh world to come.

The woods surrounding my teenage home became ankle-deep in snowdrops during those cold, interminable weeks when Christmas seemed a fading memory and the Easter holidays a very distant goal. The rough mat of ivy and damp leaf mould would gradually disappear under the mass of white and mint green mini-spears. Where the tree trunks were a little less tightly packed, it was hard to move without crushing the flowers, though if you ventured through they recovered at once, closing behind to keep the secret safe. Snowdrops are so integral a part of a wood in early spring that the stripped trees seem upheld and steddled by their presence.

The little white lanterns lighting the way along the River Skell at Fountains Abbey seem so settled into the banks that it is easy to imagine the first small body of Benedictine monks seeking inspiration there – but in fact these snow-drops arrived long after the subsequent Cistercian monastery had been disbanded by order of Henry VIII. When the estate changed hands in 1845, the new landowner, Earl de Grey, ordered thousands of snowdrops to create a fashionable riverside walk, instructing the gardeners to make the plants spell out his name along the bank in a Victorian version of flower power.

Snowdrops cluster so naturally at the sites of ruined abbeys – from Anglesey Abbey outside Cambridge to Forde Abbey near Chard, from Walsingham in Norfolk to Welford Park in Berkshire – it is often assumed they were planted by monks to symbolise purity of heart, body and soul, and to come into flower for Candlemas at the beginning of February. As the changing climate means that snowdrop opening times have varied considerably within living memory – from late February in the 1950s to early January in the 2010s – they are not the most accurate gauge for any calendar. Whether these plants came over with Christianity from continental Europe is also rather doubtful, as there is little evidence for them being spotted in Britain before the late sixteenth century. The great poets of Queen Elizabeth I's reign, Shakespeare, Spenser and Sidney, omitted snowdrops from their seasonal parades of flowers because they had yet to become established favourites of an English spring. An unmistakable illustration of a snowdrop appears as a 'bulbous violet' in John Gerard's *Herball* in 1597, but it is only identified as a 'snow drop' in the later edition of 1633. This doesn't mean that snowdrops are anything other than perfect coverings for roofless abbey floors, only that they may have found their 'natural' habitat in Britain some years after the great buildings were destroyed.

The drifts now widely regarded as the first sure sign of spring probably spread from carefully tended gardens to naturalise in congenial sites across England, Wales, Ireland and Scotland: once a few bulbs have taken root, they soon increase, through growing more bulblets on the bulbs or

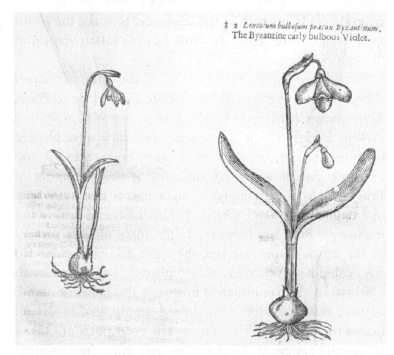

‡ 2 *Leucoium bulbosum præcox Byzantinum.*
The Byzantine early bulbous Violet.

*The 'bulbous violet' is said to be known colloquially as a 'snowdrop',
in Johnson's new edition of Gerard's* Herball, *1633.*

as the seed is blown from the flowers. Though still regarded
primarily as garden plants in the later eighteenth century,
awareness that they might also be growing wild in Britain
was beginning to take root. William Withering, whose
best-selling if unimaginatively entitled *An Arrangement of
British Plants* kept expanding into fatter editions as the craze
for botany grew, was evidently puzzled by the spread of
snowdrops. When it first appeared in 1776, they were known
to be growing wild in Gloucestershire; by 1818, there were
reports of 'Fair Maids of February' at the foot of the Malvern
Hills, near Cirencester, on the banks of the Tees, in

Barnstaple, Laxfield, Kirkstall Abbey and as far north as Scot's Wood Dean in Northumberland. A contributor from Yorkshire spotted snowdrops on the banks of the Skell – so perhaps Earl de Grey's gardeners were working with bulbs already growing wild further along the river. Withering was clear that many of the clumps were thriving 'where no traces of any buildings or gardens are to be found' – in other words, growing wild in remote places. How they got there remains a mystery, but reminds us that an apparent increase in wild flowers can sometimes reflect an increase in people spotting and recording wild flowers.

Although the bright white outcrops were increasing, it still took time for snowdrops to settle into the fertile ground of British culture. When James Thomson included snowdrops in his long poem *The Seasons*, which remained very popular throughout the eighteenth century, he presented it as the first flower of spring, leading a train of other cultivated blooms. Snowdrops still retained a somewhat exotic public image when the Regency began, judging from their inclusion in Robert Thornton's hugely extravagant *The Temple of Flora* of 1812, a grandly illustrated botanical book designed to celebrate the great Linnaeus and the world's most highly valued plants. Many of the plates feature flowers that astonished British gardeners when they arrived from far-flung lands, such as the 'Blue Egyptian Water Lily' or the 'Night-Blowing Cereus', with its circle of orange rays standing proud against a moonlit night. But here, too, ahead of the passionflowers and 'Indian Bean', is a portrait of a snowdrop so magnified by the perspective that it towers over a snow-covered landscape, accentuating the wonder of a tiny flower that rushes in where the larger but less hardy fear to spread.

The pure white flower fated to bloom only in the coldest months, leaving so little trace of its existence during the rest of the year, inspired Romantic poets such as Charlotte Smith and Mary Robinson to see it as an emblem of unjust suffering or forgotten innocence. Snowdrops, whose white petals, intricately tricked out in dashes of green, look so fragile that even the softest breath might threaten their existence, but are actually among the hardiest of plants. Despite its popular name, this is not a plant that suddenly drops at the touch of the first flake. That such a delicate form can withstand the ice storms of the earliest months remains an annual wonder, inspiring events such as a special Snowdrop Week at Altamont Gardens in County Carlow or the snowdrop festival at Cambo Castle in Fife. Snowdrop watching has become a popular winter sport – and not the most difficult to try out, given the great profusion of blooms in so many places. It does take great skill, however, to master the staggering number of different varieties, many of which are distinguished by tiny differences in the green marks on the whorls or by the size and shape of the leaves. Cambo Castle boasts some 350 different varieties, while the National Collection at Leighton Buzzard runs to over 900. At the famous rococo garden at Painswick near Stroud, snowdrops spill out in tiers, like small waterfalls frozen mid-stream. As visitors are often frozen too, there are indoor exhibitions of botanical art for those who prefer their snowdrops less chilled.

Some of the snowdrops now flourishing in the landscaped gardens of Britain and Ireland were brought home by military personnel returning from the Crimean War. The soldiers found in the steep valleys of Crimea not only cannons to the right of them, cannons to the left, but also unfamiliar species of snowdrops. These were *Galanthus plicatus*, quite similar to the species by then well known in Britain (*Galanthus nivalis*) but distinguished by the pleats along their leaves – hence their name, *plicatus* or 'pleated'. Buffeted by icy winds and blizzards, these small, white flowers must have spoken to some of the frost-bitten soldiers more clearly than the gunfire and bombardment.

Any men hailing from the north of England may have been less inclined to take the Crimean snowdrops home. In his wonderful encyclopaedia of plantlore, *Flora Britannica*, Richard Mabey reports that in Northumberland, Westmorland and Hampshire, snowdrops were viewed as 'death-tokens'; it was thought especially unlucky to take a single flower indoors. The Victorian writer and publisher, Samuel Partridge, may have known of this superstition when he published an illustrated story entitled *Snowdrops*, in which a little boy called Charlie plants a bulb for his invalid sister Meg and waits impatiently through the winter for a flower to appear. When at last the snowdrop opens, he takes it to Meg and that very night she dies. This may strike modern parents as an odd choice for small children, but it does reveal how the same flower can generate quite different meanings. Far from being a cautionary tale about the dangers of snowdrops, this prettily illustrated Victorian book is an attempt to reconcile young readers to the loss of a close family member – a calamity all too common in the days before penicillin and modern medicine. Though early in the book, when still only bulbs, the snowdrops express considerable annoyance at being pressed into the dark earth, they learn the virtue of patience as they eventually rise into the light. Rather than being sinister 'death-tokens', these consoling snowdrops belong with 'the bright angels' who carry Meg to 'the fair country'. Partridge's Victorian children's book is in sharp contrast with A. D. Miller's recent novel of the same title, with its grim depiction of contemporary Moscow, where the bodies of the homeless or victims of crime become visible as the winter snow melts, and are known colloquially as 'snow-drops'.

The botanical name for the snowdrop family is inspired by the signature whiteness of the flower, but not by its resemblance to snow. *Galanthus* derives from the Greek words for 'flower' (*anthus*) and 'milk' (*gala*). This is just the flower, it seems, to nourish the new-born year. Linnaeus, who chose the botanical name, was also conscious of its snowiness, calling the commonest species *Galanthus nivalis*, or milk flower of the snow, making it more of an ice cream or smoothie than a warm nourishing drink. It is not a good idea to attempt drinking or eating snowdrops, however, as they are rather toxic and likely to make you very sick. Charles Darwin's grandfather, the irrepressible scientist and poet Erasmus Darwin, was bold enough to boil the roots and see what they tasted like. Although he decided they were rather insipid, he seems to have suffered no ill effects judging by his suggestion that they would make a good thickener for soups and sauces.

Snowdrops have been credited with quite magical properties, if recent suggestions that these were the divine 'moly' of Homer's ancient epic are correct. This may be the flower that the beautiful Greek god Hermes gave to Odysseus to protect him from Circe's dangerous spells. On his long journey home from the Trojan War, Odysseus had landed on Circe's island where his ravenous men were turned into pigs. The only way to rescue them was to confront her in her castle, while avoiding a similar fate himself. The way to accomplish this, according to Hermes, was with a small white flower even more powerful than Circe. Odysseus wisely follows Hermes' advice and, fortified with moly, startles Circe into submission and so frees his men from their swinish form.

Homer's poems, every bit as hardy as these flowers, have survived more than twenty-eight centuries and show no sign of going extinct. Hermes' small white magic flower might suggest different species to readers in different times or regions, but for those familiar with the snowdrop, it seems a plausible candidate. Since the plant is now known to contain galantamine, which can act as an antidote to certain poisons or potions, clinical pharmacologists Andreas Plaitakis and Roger Duvoisin proposed in 1983 that it might well have been the original moly. The flowers are native to the area around the Aegean and Turkey, where Odysseus was making his long voyage home, so Homer and his early audiences would perhaps have had no difficulty in imagining their wily hero brandishing a snowdrop. There are arguments against the identification, though, not least because Odysseus needs help from the gods to lift the little plant. Anyone who has snowdrops in the garden knows how easy it is to uproot the bulbs. The most effective way of increasing their numbers is to lift and split the clumps and settle them in another bed. In a poem where sailors turn so easily into pigs and gods appear willy-nilly, literal meanings may not be the most illuminating, after all. As Odysseus' milk flower restores his men to humanity, it offers a metaphor for the possibility of positive transformations and hope for human nature. And if a small flower can sometimes be more effective than brute force, Odysseus' story still has much to say to those in power, though he did take his sword into Circe's home as well as relying on the pretty flower.

The ancient Greek heritage of *Galanthus nivalis* suits the World Heritage site of Stowe Gardens in Buckinghamshire. Here among the neoclassical temples and Elysian fields designed many centuries after Homer told the story of Odysseus, the colourless flowers form fitting cascades. From early January, the Galanthophiles arrive, braving sub-zero temperatures to marvel at the whitening slopes. In 2018, the gardens attracted virtual visitors too, as the Stowe-drop saga was aired. The electronic diary of Stowe's snowdrop family began on 3 January, with the appearance of the first tiny shoots. A single white flower-head emerged two days later, but by 15 January the original clump had 'some neighbours'. Within another week, more clumps were gathering, and by the end of the month King Alfred, the Black Prince, John Hampden and the other busts in the Temple of Worthies could gaze in satisfaction over Stowe's new white carpets. This simple photographic record encourages visitors to take a trip to see the real flowers and enables those who are not able to do so to share in the spring display. It is a very undemanding soap, reassuringly familiar and almost certain to be back next year. Online viewers may also be inspired to observe the little dramas unfolding in their own gardens. Though on a rather smaller scale than the acres of grand designs at Stowe, any plot planted with bulbs can host similar natural tales. Some gardens in January 2018 saw very early narcissi popping out beside the snowdrops, as if to add an extra storyline to the usual botanical narrative and to whet the appetite for what else might be about to unfold.

Bank of Primroses and Blackthorn. *Victorian artist William 'Bird's Nest' Hunt became famous for capturing the spring in brilliantly realistic watercolours.*

Primroses

I always felt rather bad about flattening primroses. I would never have trodden on the tufts of pale flowers when I found them in the woods, popping up from their plates of leaves like slivers of chilled butter, but so abundant were they, I felt no qualms about picking one or two from different plants and taking them home to press. Of all flowers, primroses were the best for pressing, because they kept a perfect five-petal shape and preserved their natural colour. Pale lemon with a splash of ginger, the primrose has a translucency that puts other flowers in the shade. Dirty-blonde hazel catkins look decidedly drab if they happen to open above a bank of gleaming wild primroses. After a few months between tissues and the pages of a heavy book, the primroses emerged, wafer-thin, delicately coloured and, unlike their contemporaries, saved for posterity. I would comfort myself that, far from cutting short their brief lives, I was immortalising a few blooms each year. It was only at the moment when the remorseless pages clamped down on the unsuspecting flowers that I would hesitate, wondering whether they might be better kept in a small glass vase or, better still, left to bloom and fade on the slim stalks sprouting from the woodland floor. A bunch of primroses for Mother's Day, depending on the weather and the timing of the festival, was part of the annual cycle, as natural as the sudden showers or the flashes of sun that made the wet petals shine. The pressed primroses, on the other hand, would reappear unseasonally, during the muddy brown months of November and December, as precious relics of the spring. I liked dismantling my great pile of books, opening the pages and gently peeling back the paper to reveal the pressed primrose, a purer yellow than the slightly foxed old pages, with a deep golden flame of a heart. My dried flowers would then be arranged around the short poems I copied in my best calligraphy ready to be given away as Christmas presents. The primroses surrounding Tennyson's 'The Brook', which my sister has loyally treasured ever since, must now be forty years old. The plant which produced the flowers may still be sending out similar rosettes every spring, to claim the annual gold medal for being itself.

The leaves of the common primrose complement the thin, semi-transparent lemon petals: ebullient and abundant, succulent and edge-curled, they unfurl in dense layers of deep green. It is as if the most delicate flowers need the strongest protection in order to venture out. Even the tiniest seedlings are surprisingly robust, their immaculately formed miniature leaves shooting out in all directions. These tiny green ears, slowly unfolding like new-born skin, seem to be listening out for the spring. They spread into elliptical openness, but the crinkly leaves, once fully grown and scored by pale veins, converge on a vanishing point beneath a cluster of slim stems. The round mounds of moss-stitched green make an emerald feather quilt to wrap the hedge bottoms and the roots of trees. The blanket becomes studded with brightly coloured buttons as the strong leaves come under the sway of their golden crowns. Strange that if the colours should be inverted and the green leaves go yellow in their turn, the primrose picture of health turns sallow, sickly and inauspicious. Though primroses survive in most kinds of soil, they need moisture to thrive. An early drought will sap their vigour, transforming those succulent leaves into shrivelled brown shadows. This is why they often do best in wooded areas, where the leaf mould holds the moisture while the light streams in until dammed by the thick summer foliage of mature trees.

In Devon, the county of primroses, mild moist weather means that the flowers flourish in more exposed sites, forming drifts of colour from the hillsides around Woolacombe Bay to the valleys near Newton Abbot further south. From late March and into April, the steep banks beside the roads are cushioned in yellow and green, as if the entire county has been reupholstered in honour of spring. So plentiful were these wild flowers that for more than a century, Wiggins Teape, the company that ran the large paper mill at Ivybridge on the edge of Dartmoor, would dispatch thousands of small boxes packed with Devon primroses as a gift to all their customers every spring. The goodwill gesture was only discontinued in 1989, in response to growing concerns about the depletion of local primroses. In fact, the primrose decline probably had more to do with intensive agriculture than over-picking, though climate change has also had unpredictable effects on many native species. On the far side of England, Oliver Rackham was noting a drastic decline in primroses among the Cambridgeshire woods and attributing it to the changes in farming, greenhouse gas emissions and rising temperatures. Devon remains proud of its signature flower, perhaps appreciating it all the more amid fears that it might become the 'uncommon' primrose. The old practice of primrose picking has now been recast as a cherished local tradition and was celebrated in a new primrose festival at Ivybridge on Mother's Day 2017.

Some wild primroses flower in pale raspberry, peaches and cream, or almond white, as well as every shade of apple juice and white wine, though the common primrose, *Primula vulgaris*, usually appears in the colour that has become widely known as 'primrose yellow'. Much harder to spot than the common primrose is the tiny Scottish primrose, *Primula scotica*, which flies its purple flowers from a small mastlike stalk. Although common primroses grow in many parts of Scotland, the Scottish primrose is only found in the cool coastal areas of Caithness, Sutherland and Orkney, doing best where the local cattle are not too greedy and keep the grass at the optimum level for primrose well-being. In the Pennines, the becks and limestone crevices are home to their own local primrose, the graceful bird's eye, *Primula farinose*, with its beady-eyed, pink-hearted ring of petals.

A great favourite, wherever it survives, is the *Primula veris* or cowslip, whose flower clusters hang from the stalk like little green bags bursting with bright yellow icing. In Somerset, these distinctive flower-heads suggested the celestial names of 'golden drops', 'bunch of keys', 'St Peter's keys' or 'herb Peter'; in other regions, as Geoffrey Grigson discovered, the local names were rather more down-to-earth, though equally vivid, variations on cow dung: 'cooslop' in Lincolnshire, 'cowflop' or 'cow slop' in Devon, 'cow slap' in Hertfordshire and 'cow strupple' across the north of England. Despite what might seem rather off-putting names to would-be pickers, the flower-heads were often snipped off by children, suspended on ribbons and then tied into little cowslip balls or chaplets for bridesmaids. While they seemed to be in severe decline in many areas in the 1990s, they were flowering in abundance in 2018 along the banks of the M40. The speed at which most people could see them meant that it wasn't easy to appreciate the botanical details that inspired their older, rural names, but the mass of little cluster-heads gave the motorway cuttings a most unusual, lemon sherbet sheen.

For Mary Russell Mitford, whose recollections of her rural home near Reading became the enduring favourite, *Our Village*, making cowslip balls was a pleasure of May. Far more exciting, however, was the start of spring itself, which was marked by 'the first primrose'. Every March, she would walk her dog through the Berkshire lanes in search of primroses. Her walks were hardly worth having until she had spotted the first yellow face peeping through the undergrowth. By April, she would set off for her favourite copse, confident despite the harshest winter in the prospect of finding the golden centre of what she called the 'primrosy' season. What made this particular bank of primroses so pleasing was not the primroses alone but 'the natural mosaic' of 'ground-ivy, with its lilac blossoms and the subdued tint of its purplish leaves, those rich mosses, those enamelled wild hyacinths, those spotted arums, and above all those wreaths of ivy' which linked all the spring flowers into one harmonious whole. The fragrance of the primroses in the sheltered copse was wafting in the spring. And here was the first butterfly of the year, too, exactly matching the colour of the petals as it alighted on a primrose clump.

Primroses are nurseries to some of Britain's rarest butterflies, such as the Duke of Burgundy, which hides its eggs in the furry underside of their foliage, but the one spotted by Mary Russell Mitford was almost certainly a male brimstone butterfly, once known as a 'primrose'. As these light yellow brimstones suck the early nectar from a sequence of primroses, it is easy to think for a moment that some of the flowers have taken flight. In John William Inchbold's strikingly realistic painting of 1855, *A Study, in March*, the brown earth bank below the brightly lit sycamore trunk includes a well-observed primrose, with two flowers in full bloom, other buds just opening and, in similar hue very close by, a small brimstone butterfly.

The close association between the sulphur-coloured flowers and butterflies is plain in John Clare's poem 'Primroses', which celebrates the appearance of 'pale brimstone primroses like patches o' flame' in the dark wood during spring. While primroses offer crucial sustenance to butterflies on bright, but chilly, days from March to May, their visitors are busy transferring pollen from plant to plant. If you stare into the centre of a primrose, you will soon discover that not all the flowers are identical. Some have a small green bull's eye, while others centre on a yellow pom-pom. These are known respectively as 'pin-eyed' and 'thrum-eyed' primroses, the 'pin' being the stigma, protruding above the anthers in some plants, the 'thrum' being a cluster of anthers, which forms at the top of the stamen, well above the stigma, in the other kind. (A 'thrum' was originally a weaving term for a fringe of loose threads and sometimes used to refer to a mop head, which is what these small, yellow clusters resemble.) A brimstone butterfly with a long proboscis will reach deep into a primrose for a sip of nectar and, in doing so, collect pollen from the anthers, which then drops at the next fly-through flower stop. The thrum-eyed primroses pollinate the pin-eyed and vice versa, because of the mutually advantageous arrangements of anthers, stamens and stigma at different heights within the two kinds of flower. This remarkable compatibility was first explained by Charles Darwin, who calmly set to work on different kinds of primrose while the controversy raged over his book *On the Origin of Species*. Looking back on his life some years later, he confessed to harbouring a very soft spot for the primrose, since nothing in his scientific life had given him more satisfaction than 'making out the meaning of the structure of these plants'. If many of his contemporaries delighted in the primrose's colour, form and clusters of associations, for Darwin it was the internal organs that mattered.

Although primroses have now been overtaken by snowdrops as the first flowers of an English spring, they have kept their old name, 'prima rosa' or 'first rose'. These were not only the flowers that traditionally appeared earlier in the year than other spring blooms, as the bleak winter finally went into retreat, but 'prime' in the sense of excellence: the primrose is the first and foremost flower. This may explain why John Milton called this plant the 'rathe primrose' – to emphasise that this is not only an early riser, since rathe means 'early', but also a paragon – like the young, drowned poet lamented in his elegy, *Lycidas*. Milton knew that youth was regarded as the primrose-time of life and that primroses could suddenly be cut off in their prime, unjust and inexplicable as this seemed to those left to mourn. This was the flower 'that forsaken dies', like the young poet snatched from the earth before he could reach fruition. In an early draft of the poem, the primrose in *Lycidas* was 'unwedded' rather than 'forsaken', emphasising the pale, delicate flower's unhappy associations with those who outlive their beloved and then die prematurely of grief. Contemporary plantlore linked primroses to those who perished too early and whose promise was unfulfilled. Milton was recalling Shakespeare's related reference in *The Winter's Tale* to the 'pale primroses / That die unmarried ere they can behold / Phoebus in his strength', which very concisely fused observation of spring flowering habits with concerns about 'green sickness' or chlorosis, a condition caused by iron deficiency and characterised by loss of colour, that can afflict plants and adolescent girls.

For Milton, as a Christian poet, the final choice of the word 'forsaken', rather than 'unwedded', brought thoughts of the crucifixion into Lycidas's untimely death and with them the hope of resurrection. The primrose's habit of flowering in March and April has helped to strengthen its association with Easter. This was – and is – a flower picked to decorate country churches for Easter Sunday after six weeks of Lenten austerity. Suddenly fonts, pulpits, altars and porches are covered in green moss and primroses. When Francis Kilvert served as a curate in the rural parish of Clyro, near Hay-on-Wye, he was amazed by the Easter primrose decorations in the church and even more so by the local custom of covering the graves in the churchyard with flowers. As he walked through the churchyard under the moonlight, the decorated graves looked 'as if people had laid down to sleep for the night out of doors, ready dressed to rise early on Easter morning'. Kilvert himself rose early on Easter Sunday and, as he gathered primroses from the frosty banks by the mill pond at six o'clock in the morning, he heard the first cuckoo of the year. Contemporary artists often saw primroses as inseparable from birds, or at least from eggs, painting the plants beside clutches cradled in a nest of twigs and moss. The water-colourist William Henry Hunt became known as 'Bird's Nest' Hunt because of his highly accurate depictions of nests and primroses, which prompted a host of imitators. In the Victorian art world, primroses began to usurp bacon as the standard accompaniment for eggs. To modern eyes, the removal of nests full of bird's eggs and entire uprooted primrose plants to an artist's studio may seem a strange way to celebrate the natural world, but for many of our great-great-great-grandparents these were devotional images of new life and signs of God's creation.

It may have been the flower's old association with primacy, however, that gave it special appeal to Benjamin Disraeli, in his determined ambition to become a Member of Parliament and, eventually, prime minister. The common wild flower formed an unlikely bond between Disraeli and his sovereign, because Queen Victoria, grand empress of half the world, could still be moved by simple pleasures and romantic gestures. She would send her favourite statesman Isle of Wight primroses from Osborne House. (The flowers were so prolific in the Isle of Wight that when John Keats visited in 1817, he thought it should be renamed 'Primrose Island'.) When Disraeli died in April 1881, in place of a grand funeral wreath the Queen sent a circlet of primroses for the cortège, with a handwritten card saying 'His Favourite

Flowers'. Two years after Disraeli's death, Henry Drummond Wolff, MP for Portsmouth, was surprised on his arrival in Westminster to be offered a primrose by a cloakroom attendant, but even more so when he discovered that all the Tory gentlemen in the House of Commons were wearing primroses in their lapels in his memory. A few months later, in November 1883, Henry Wolff met Lord Randolph Churchill, Sir Alfred Slade and Sir John Gorst at the Carlton Club, and they agreed to promote an annual primrose day on 19 April and to found the Primrose League in memory of the late prime minister, Lord Beaconsfield.

Garlands for Disraeli: crowds gather in Westminster to remember the late prime minister and his ideals of 'religion, loyalty and patriotism' on Primrose Day 1886.

The Primrose League upheld Disraeli's principles of 'Religion, Loyalty and Patriotism', which soon became 'Empire and Liberty' and saw branches springing up across the UK. Within four years, it boasted over half a million primrose-wearing members; by 1891, the number doubled again. Members sported little yellow badges and gathered for tea parties, summer fetes and musical recitals. Since the yellow colour of the primrose had old associations with anti-Semitism in medieval Europe and the Islamic world, this new mark of respect was helping to transform ancient prejudice into affectionate celebration. Women, stirred by the possibility of political involvement, but of less radical inclinations than the suffragettes, flocked to join the Primrose League, though the new Primrose Club in 4 Park Place, St James's, was, of course, open only to 'gentlemen holding Conservative Principles'. The pretty if not very long-lived, yellow-bordered, floral-cornered *Primrose Magazine* launched in 1887, with its mix of political comment ('Fair Erin: Her Beauty and her Troubles'), imperial travelogues ('A Jaunt through Burmah'), poems about primroses and copious advertisements for Hair Fluids, India Carpets, Railway Insurance or Turkish Baths 'in your own room' aimed at a mixed readership of ladies and gentlemen.

The newly politicised identity of this flower was some-what unsettled by the ascendancy of Lord Archibald Primrose, 5th Earl of Rosebery, who succeeded Gladstone as the Liberal prime minister in 1894. His family coat of arms had long sported stylised primroses and even the jockeys who rode their prize racehorses were equipped with jerseys striped in primrose pink and yellow. As he held power for little more than a year, his Liberal policies were not popular enough to uproot Primrose Day, which was now firmly established as an annual festival. One of the earliest reels of Pathé news captures Primrose Day in 1916. Disraeli's statue at Westminster, adorned with a huge wreath of primroses, rises above a crowd of young soldiers kitted out for the First World War. It is just possible to make out in the grainy grey footage a flower-seller, not unlike Queen Victoria in shape and stature, offering a tray of small bunches to the tall, young men all around. The primroses made a touch of soft colour on their moss-coloured uniforms, as they embarked for northern France.

The primrose is at once the freshest, purest of spring flowers and yet its strong associations with youth mean hazard as well as hope. Perhaps the way those healthy emerald leaves so easily turn a thick, sickly yellow, as if in mockery of the delicate flowers, have always made people wary of the primrose's capacity to fall from grace. Shakespeare knew there was nothing prim about these flowers and understood the dangerous appeal of their innocent appearance. The drunken Porter at the gates of the castle in *Macbeth* expects to let in 'some of all professions that go the primrose way to th'everlasting bonfire'. Whether this was the common primrose or the Scottish primrose is not specified: either purple or pale gold might be suitable flowers for the royal castle of Dunsinane. The porter's dark jokes are reminiscent of another of Shakespeare's great tragedies, *Hamlet*, where Ophelia rebukes her brother for offering advice about her personal conduct with a warning against hypocrisy: he should think twice about pointing her towards 'the steep and thorny way to heaven', while he himself 'the primrose path of dalliance treads'. Whether lining a path or a way, primroses are so enticing and seemingly innocent that no one suspects them of leading very far from virtue

– until it is too late to return. Oliver Goldsmith may have had this in mind when he called his hapless Vicar of Wakefield Charles Primrose, giving him sermons on the sacred bonds of matrimony, while his daughter, Olivia, was catching the eye of a predatory local squire. In the end, things turn out relatively well, but the fragility of primroses is plain enough.

The irresistible and multifaceted appeal of this flower is obvious wherever it grows. More butter-coloured than the buttercup, the primrose inspired the Norwegian cheese-maker and philanthropist, Olaf Kavli, whose experiments with dairy products eventually gave rise to the world's first soft, spreadable, long-life cheese. The pale yellow whole-someness of Kavli's creamy cheese made 'Primula' the perfect name. The common primrose, *Primula vulgaris*, again linked to cows in Norwegian as *kusymre*, grows wild across Norway and the Faroe Islands. Kavli grew up in a village on the Romsdal peninsula, where the damp temperate climate and fertile ground encourages both dairy farming and primroses. Once his revolutionary cheese was ready for mass produc-tion in 1924, he set up a factory in the port of Bergen and was soon exporting to the rest of Scandinavia, Britain, Europe and eventually the world. The pastoral origins of these foil-wrapped half-moons and tubes of cheese were underlined by the cheerful, five-petal, cheddar-coloured flower, whose name appeared on every pack. And, if the name wasn't enough, the beaming young woman with flaxen hair, scarlet heart on her apron and an armful of little yellow flowers, was enough to reassure anyone that nature was simply waiting to be unwrapped.

Real primroses once furnished British tables in the form of primrose pottage tea, which may well sound more attractive than it tasted. As these plants have natural sedative properties, people probably slept well on it nevertheless. The petals infused in water made a brew that was recommended for calming nerves and hysteria, while the desiccated roots were prescribed for headaches. The leaves, harvested before they became too substantial, could be added to salads and an aromatic oil extracted from the flowers and roots (though the primrose oil now commonly available as a herbal remedy comes from the evening primrose, a completely separate native American species, *Oenothera biennis*). By far the prettiest culinary deployment of primroses has come with the contemporary fashion for imaginative cake decoration. A handful of flowers, coated in egg white and sugar and scattered over smooth, snow-white icing, gives the effect of primroses glimpsed at dawn on a frosty spring morning – far too delicate to cut up and eat.

The primrose inspires a perennial desire for preservation, whether through sugar-coating, silver foil, flower pressing, painting, poetry or conservation. In fact these plants are often surprisingly long-lived and, if undisturbed by spades or changing conditions, will keep on flowering every spring for many decades. And yet, these familiar ground-breaking flowers, whose beauty has been admired so often that the annual astonishment is astonishing in itself, will still disappear just as surely every year, leaving us to lament that their brief lives are not just that little bit longer.

Daffodils make a seasonal challenge to the white rose as the flower of York in a British Railways poster of 1950.

Daffodils

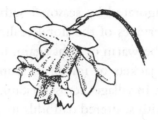

At first, you hardly see them: slim, green tips, barely distinguishable from the dew-drenched grass. Some, it's true, are not quite so slim – here and there are pale swellings, hardly enough to alter the outline, but firmer, somehow, and more promising. Day by day, they become less reticent and, though some stay hidden among the clumps of chaotic dead grass, others stand tall and straight, ready to meet the sun when it may choose to appear. Their heads begin to tilt, as if still too shy to look you straight in the eye, but almost at once their astonishing secret is out – in a flash of yellow and a silent, spectacular chorus of trumpets. Daffodils herald the spring with vim and verve, seizing the limelight from the more diminutive early risers. Brighter than snowdrops, taller than celandines or aconites, daffodils instantly command attention. Undeterred by heavy rain, they emerge from the shower dripping wet and glossier than before. A hard frost is more of a challenge, forcing their open faces downwards and turning their pliant stems into stiff, green, frozen arches. Late snow will bow them to the ground, stalks collapsed in all directions, heads buried under the weight of wet whiteness. But only prolonged falls of heavy snow can utterly defeat a company of daffodils: when the sun returns and the ice retreats the yellow stars will generally be out again. Daffodils, upright and bright, seem invigorated by seasonal setbacks. They form the emergency services of the floral realm. Just as it seems that it will never be warm or light again, huge teams appear in their high-vis jackets to rescue us from the effects of a prolonged winter. In villages across Britain, worn-out verges begin to gleam with scattered daffodils in early spring, until the roadside is ablaze in a luminous citric glow, often outshining the sparse streetlights. As the chilly dawn begins to break a little earlier, these seasonal signs remind yawning drivers and shivering teenagers that things are not, after all, quite as dark as they have been and that brighter days are coming.

A few miles along the road from our house is a field of daffodils, planted some years ago by a young family in deep mourning for their mother, who died of cancer at the age of forty-five. The Field of Hope becomes brighter as the years pass because of the daffodils' natural tendency to spread and increase. There are vast golden Fields of Hope in Sefton Park in Liverpool, beside the Manchester Business Park and along the banks of the Tweed, and much smaller, but no less hopeful, displays at the primary school in Madden, County Armagh, or on the Sleat peninsula in Skye. The Marie Curie cancer charity began selling fresh daffodils in March 1986 to raise money for the nurses who provide care for cancer patients and support for their families. The flower initiative was so successful that what has become the annual Great Daffodil Appeal now distributes boxes and boxes of yellow daffodil-shaped badges. These little bursts of sunshine often brighten up some quite unlikely lapels. Rail passengers alighting at Glasgow Central on 16 March 2018 were dazzled by the new Garden of Light – an illuminated display of vibrant daffodils in the concourse. Though no one could have doubts about the good of the cause, the success of the daffodil campaign has occasionally provoked some criticism as the 1st of March seemed to become known as Marie Curie Day. As daffodils had already been worn on this day, which is St David's Day, for many years in celebration of Wales, the charity felt moved to issue an apology for the accidental usurpation. In March 2018 the rapprochement – or renewed rivalry – was clear for all to see when the charity's ten-foot-tall light-up daffodil installation was unveiled in the grounds of Caernarfon Castle.

The Welsh name for daffodils is *cennin Pedr*, or Peter's leeks, because the flowers begin to open at about the time of the feast of 'Peter's Chair', on 22 February, as Nick Groom explains in his book *The Seasons*. They are Peter's 'leeks' because their long stalks and bulbs resemble the much older Welsh symbol, and botanical relation, the leek. (When Shakespeare wanted to demonstrate the national feeling of the very Welsh Captain Fluellen in *Henry V*, he gave him a leek as a badge of honour rather than a daffodil.) Daffodil sounds like just the flower for Saint David, though, the patron saint of Wales, whose cathedral is in Pembrokeshire and whose feast-day is celebrated on 1 March, a week later than St Peter's Chair, when even more daffodils are in bloom. A single flower sported on a cap or jacket may stir the heart, but daffodils are best in massed choirs; hence, perhaps, their natural affinity with the Welsh.

In fact, the word *daffodil* derives from the Greek asphodel, a tall white plant, whose six-petal star flowers are not unlike some kinds of daffodil. 'Daffodil' does not have quite the stately ring of 'asphodel', which, as is often the way with fabulous flowers, soon came down to earth as it hybridised with British culture. The Greek word quickly naturalised as 'affodil', 'daffodil', 'daffadilly' and 'daffy down dilly' to blend in with popular English names such as 'Easter lily', 'Lent lily', 'Lenty cups', or the more rustic 'goose flops', 'sun-bonnets' and 'butter and eggs'. The classical heritage of the asphodel is rather less homely. In ancient myth, pale asphodels dwell in the meadows of the dead, beneath the sway of Queen Proserpine (or Persephone, as she was known in Greece), who lives there for half the year. Proserpine was swept away by Pluto, god of the underworld, when he spied her picking spring flowers in the fields of Enna. Her underworld fields of asphodels are the pale shadows of her former life on earth. The daffodil's annual habit of disappearing entirely for many months and then bursting up with renewed vigour fits well with ancient vegetation myths of disappearance into the underworld and reappearance in the summer months. Its April flowering in the northern hemisphere also fits the Christian festival of resurrection at Easter.

Salvador Dali's well-known painting, *Metamorphosis of Narcissus*, draws on the myth of Proserpine to challenge any comfortable sense of the separation between life and death, winter and summer. Surrealism, no respecter of convention, blends the ancient and contemporary, classic and kitsch, myth and – er – myth into unreal realities. Here memories of Proserpine play into the ancient story of Narcissus, who gazed with such passionate admiration at his own reflection that he gradually pined away, leaving only a white flower with a yellow centre in his place. Dali offers multiple reflections on Narcissus: as a figure kneeling in a pool; as a young man, set apart from the rest of the party on a pedestal at the centre of a chessboard; and as a single narcissus flower, sprouting unsettlingly from a cracked egg, which is poised between a vast finger and thumb. The flower resembles an asphodel as well as a white daffodil, while the sculptural hand is mirrored by the kneeling figure, turning it into a monstrous brown bulb or a giant hand, rising from the dark pool as if from the underworld, to snatch Narcissus or hatch new life.

In Ovid's tale of Narcissus, the beautiful young man, who plays with the affections of so many adoring admirers, is punished by falling so deeply in love with his own image that he kisses the water repeatedly, before finally realising that his best beloved is nothing more than an illusion. His self-obsession is all the more cruel because of the smitten but spurned figure of Echo, the nymph who retreats in anguish and dwindles to a mere voice. All that remains of Narcissus is a single daffodil beside a pool, which symbolises the vanity of a young man so transfixed by his own beauty that he ignores everyone else. The Victorian artist, John Waterhouse, made the point emphatically clear in his painting of Narcissus by including an equally beautiful Echo, trying to catch his attention in very revealing drapery – but to no avail: the young man outstretched beside the mirror-like pool, with a white narcissus at his feet, has eyes only for himself. It is strange that such an unhappy myth should have been foisted on to daffodils.

Echo fails to divert Narcissus' attention in J. W. Waterhouse's Echo and Narcissus, *1903.*

William Wordsworth knew very well that the wild, lemon-petalled daffodils of the Lake District, *Narcissus pseudonarcissus*, rarely appear in isolation. Those exploring the area in April may still see them clustering in the grassy churchyard at Cartmel Fell or scattered along the rocky banks of the River Duddon. They are more colourful than the *Narcissus poeticus* depicted by Waterhouse, but paler than many of the very yellow daffodils planted in modern parks, gardens and roadside verges: a softer colour is often the sign of a hardier, wild species. These are the diminutive daffodils that once grew wild across much of Britain, but now rarely appear in great numbers outside the Lakes, the Black Mountains, parts of Devon, Yorkshire, the South Downs and the border of Gloucestershire and Herefordshire. In the woods at Gowbarrow on the banks of Ullswater, the daffodils which inspired Wordsworth's most famous poem still bloom each year. Unlike the single narcissus of myth, they were – and are – most striking because of their profusion. Wordsworth came across these daffodils during a walk with his sister Dorothy on Maundy Thursday, 1802. At first they spotted a few by the water's edge and wondered whether the seeds had floated ashore from some clump on the far side of the lake; walking on, they found themselves before a broad border of waving flowers, which ran along the shore like a golden road. Dorothy Wordsworth was so elated that she jotted down the experience in her journal to capture and keep the joy of it:

> I never saw daffodils so beautiful they grew among the mossy stones about & about them, some rested their heads upon these stones as on a pillow for weariness & the rest tossed and reeled & danced & seemed as if they verily laughed with the wind that blew upon them over the Lake, they looked so gay ever glancing ever changing.

William's poem, written more than two years later and inspired as much by Dorothy as by the daffodils themselves, is all about shared experience and release from narcissistic reflection. The famous self-image of the lonely cloud is instantly blown away by the unexpected crowd of daffodils, 'Ten thousand dancing in the breeze'. This lyric is often quoted to conjure up the image of the solitary Romantic, but the host of golden daffodils is really a retort to centuries of misplaced myth: in no time at all, the poet is celebrating the joy of joining in. The poem even includes lines composed by Wordsworth's wife Mary, who was not even there for the walk along the shore, but could still share in the pleasure of the daffodils later through the Wordsworths' vivid words. Mary's contribution – 'They flash upon that inward eye, / Which is the bliss of solitude' – recognises the longer-term benefits of a startling encounter with natural beauty. The sight of a lakeside utterly submerged in waves of yellow can live on in the mind, transforming loneliness into 'bliss'.

This may help to explain why poems like Wordsworth's 'The Daffodils' have a powerfully therapeutic effect on people suffering from Alzheimer's disease, often such an isolating condition. Patients with dementia sometimes respond with remarkable lucidity to hearing a poem memorised in childhood, when other kinds of communication are evidently making no sense. The strong rhythms and uplifting sentiments of Wordsworth's 'The Daffodils', once a poem routinely memorised by schoolchildren, can surface again decades later, to give pleasure, reassurance and a renewed sense of dignity to patients deeply confused by their current condition. The Welsh poet, Gillian Clarke, captured this phenomenon in a poem of her own, 'Miracle on St David's Day', which describes an elderly Alzheimer's patient suddenly reciting Wordsworth's 'Daffodils'. He had learned it by heart years ago and, in a rare moment of recognition, remembers 'there was a music / of speech and that once he had something to say'.

In an irony of history, Dorothy Wordsworth, whose own delight in daffodils has indirectly come to the aid of dementia sufferers in the present, became a victim of the disease herself in later life. When William and Mary's daughter Dora died in 1847, Dorothy sank into utter decline as her brother, in his late seventies, withdrew into himself and wept. And yet, in the wake of Dora's death, the family planted masses of daffodils in her memory on the steep slope of their garden at Rydal Mount. Wordsworth died three years later, and Dorothy in 1855. Mary Wordsworth lived on for a few years longer to witness the daffodils in Dora's Field returning each year and to remember those she saw no more.

Across the hills from Rydal, on the slopes of Lake Coniston, is another field that turns a brilliant lemon-yellow every spring. The great Victorian artist, critic and social reformer, John Ruskin, began planting daffodils in his garden at Brantwood, after experiencing intense delight in the Alpine flower meadows he had seen on his extensive travels in Switzerland. Gazing at the waving heads of mountain daffodils, Ruskin had experienced them as a gift from God. Later, he created beautiful drawings of the same flowers and would instruct his students to pay the closest attention to them, painting their cup in 'the yellow which is its yellow, and the stalk of the green which is its green, and the white petals of creamy white, not milky white'. As if this were not challenging enough, he then insisted that these colours should be modified 'so as to make the cup look hollow and the petals bent', while still retaining the overall impression 'which is the first a child would receive from the flower, of its being a yellow, white, and green thing, with scarcely any shade in it'. Ruskin's favourite flower was *Narcissus poeticus*, with its paper-white petals, which he had seen growing abundantly in the Alps. These graceful flowers still flourish along the Harbour Walk past the daffodil meadow at Brantwood, in line with the vision of Ruskin's devoted cousin and carer, Joan Severn, who created the walk down to the lake when Ruskin was too ill for planning or planting. Although he was no longer able to see the unblinking, red-rimmed pheasant's eyes of his favourites, their heavenly scent surrounded him and took him back to healthier days.

Daffodils can now offer new kinds of help to those with Alzheimer's. The alkaloid, galantamine, found in the bulbs of daffodils, is known to inhibit the enzyme acetylcholinesterase, which is especially prevalent in patients with vascular dementia. Originally identified in 1947 in snowdrop bulbs (which explains the name, derived from the snowdrop's botanical name, *Galanthus*), galantamine was later discovered in daffodils, too. Clinical trials began in the 1990s and the compound is now extracted from the larger, more commercially viable, bulbs of daffodils for the pharmaceutical industry. Since thousands of bulbs are needed to produce even small quantities of the drug, the potential market is still uncertain, but it offers new possibilities to farmers in challenging upland regions, because the higher altitudes mean a higher concentration of galantamine than that which develops in the flat, low-level fields where commercial daffodil-growing mainly takes place. Pharmaceutical extraction is a highly specialised process, however, and no one should make the mistake of thinking that eating daffodil bulbs is beneficial. In unfortunate cases where these brown, toxic bulbs have unluckily been mistaken for onions, the consequences, though not fatal, have been decidedly unpleasant.

Most commercially grown daffodils are destined for the horticultural and cut-flower markets. Though they are symbols of Wales and the Lake District, the majority of British daffodils are farmed in the far south-west and along the east coast. In the flat fenlands around Spalding in south Lincolnshire, in the low-lying farms near Montrose in Angus and right across the Cornish peninsula, huge fields turn yellow each spring. In the Isles of Scilly, the milder climate allows the early flowering, though decidedly tender, *Narcissus tazetta* to grow, as it does in California, China, Greece, Israel, Japan and in Spain, where the species probably originates. In most of northern Europe, the winter temperatures plummet too far for the sweet-smelling, multi-headed *tazetta* to survive outside, and so it retains an air of the exotic. The weather in the Scilly Isles, noticeably kinder than in the more northerly daffodil centres, means that there are always supplies of fragrant narcissi for Mother's Day, however early in the year it falls. During the Second World War, when parks and gardens were being ploughed and planted for essential food supplies, Churchill granted special permission to the islanders to carry on growing their narcissi. The United Kingdom is the largest daffodil producer in the world, with Cornwall alone producing an annual crop of 30 million tonnes of bulbs. This is a double-season industry, because bunches of tight, chilled buds are bundled up like plastic pipes for wholesale delivery in the spring, and then the bulbs are harvested through the summer months, before being packed off around the world.

Modern daffodil farming results from one of the lesser known revolutions of the nineteenth century, which began with the experiments of pioneering horticulturalists such as William Backhouse, who came from a family of bankers and botanists in County Durham. Backhouse cultivated daffodils of a kind never seen before, such as the 'Emperor' and 'Empress', or his two-tone tetraploid superstar, 'Weardale Perfection', which rose to a height of two feet and opened its petals to a span of five inches. This was a giant of a daffodil, twice as tall as the little, wild *Narcissus pseudo-narcissus*. 'Weardale Perfection' became the grandsire of numerous hybrids and cultivars, but the very success of its progeny meant that within a few decades the original daffodil had been superseded and almost forgotten. In the late 1930s, one of its distant descendants, 'Kanchenjunga', a huge mountain peak of a flower, bred by the great Irish horticulturalist

Guy Wilson, had risen to dominate the daffodil horizons. Sixty years later, a serious hunt was on to recover any surviving 'Weardale Perfection' daffodils, and eventually a rare bulb was discovered in a garden once owned by a nurse who had worked in St John's, Wolsingham, where the very first daffodil was raised more than a century before. By now, this was more a case of 'Weardale Imperfection', but after patient work on the precious bulb, the daffodil expert David Willis succeeded in growing the veteran daffodil once again. In 2007, young bulbs were ready for planting and the vintage blooms offer a modern spin on the daffodil's age-old associations with reappearance and resurrection.

The genealogies of modern daffodils make even those of the Old Testament seem straightforward. David Willis's brilliant history of daffodils, *Yellow Fever*, is like a huge family saga, detailing hundreds of crossings and cultivars over the past two centuries, in the hands of devoted daffodil breeders in England, Ireland, Scotland and Wales, the Netherlands and, subsequently, Australia, New Zealand and the United States. There are now more than 26,000 varieties, with the number ever-increasing. To help us keep up with this proliferating array, the Daffodil Society of North America, founded in 1955, has designed a database called Daffseek, which may sound like a dating agency for gardeners, but is in fact a detailed record of every known variety. There are tall daffs, small daffs, early daffs and late, trumpet daffs with enormous cups, and daffs with tiny, pinhead eyes and circles of elegant petals. Most of these windmill flowers have six sails, but some are so puffed with frilly layers that it is hard to distinguish the corolla from the perianth. With such a bewildering array, many gardeners prefer to go for a bag of mixed bulbs and a hope-for-the-best attitude.

As different species of daffodil flower at different times, more judicious planting can ensure waves of colour to wash over brown, winter gardens. Among the first to appear is the little golden-hearted, lemon-sherbet hybrid known as 'Rijnveld's Early Sensation', which has been braving the snow and frost since the 1950s. After the vanguard of early, dwarf daffodils come the full, yellow battalions of March and April, followed by the creamy-coloured, delicately rotund, double-cupped 'Cheerfulness' narcissus or, later still, Ruskin's favourite, *Narcissus poeticus*, and the wedding petticoat daffodil oddly named 'Rose of May'. The flopping, browned-off stems of earlier flowerers are not such an attractive addition to spring beds, but leaving them intact for a few weeks to feed the bulbs ensures bigger, brighter blooms the following year. Some careful gardeners tie knots in the fading stems to minimise the unsightly sprawl, before pulling up the bulbs in May to replant in the autumn. In wilder gardens, daffodil bulbs are often left to expand underground, forming new bulblets. A large mother bulb will have more teatlike protuberances than a cow with a muddy udder. As the plants naturalise, the clumps of flower stems multiply, increasing the seeds and the extent of the display. Left to their own devices, the bulbs will spread and the flowers will seed to form abundant drifts of colour.

The vagaries of the calendar have quite an impact on daffodil growing in the British Isles, adding a certain frisson to the planning of special daffodil-centred events. For the Easter weekend of 2016, the biggest daffodil festival in Britain was due to take place in the village of Thriplow in Cambridgeshire. An unusually mild winter meant that, unfortunately, most of the daffodils had already opened, bloomed and shrivelled away before all the visitors arrived.

The poet of the moment was not Wordsworth, but Herrick – 'Faire Daffadills, we weep to see / You haste away so soone'. As this was the forty-eighth year of the festival, the organisers had made contingency plans and planted thousands of new, late-flowering bulbs the previous autumn, just in case. Even this precaution could not quite allay the disappointment of the early display, though at least Mother's Day, three weeks before, must have been a dazzling occasion. Undeterred by the disappointment, the organisers went ahead the following year and were rewarded with a fine display of daffodils in the lovely spring sunshine. In 2018 spirits were somewhat lowered again along with the temperature, which brought ice, sleet and late snowfall.

Weather is not the only obstacle to be overcome. For badgers and foxes, daffodil bulbs seem to have a special magnetism – gardeners everywhere are provoked by these nocturnal thefts. After lugging a whole sack of bulbs home, spending back-breaking hours burying their large brown treasure in the earth, comforting themselves with the promise that it will soon be turning into gold, they wake up the very next morning to find the ground studded and furrowed and the new bulbs largely gone. This is all the more irksome for a meticulous planter, who has followed the producer's advice to sink the bulbs at least six inches below the surface – no mean feat in hard, rocky or water-logged ground, especially for those without a special bulb dibber. The battle with autumnal mud and peckish predators does enhance the eventual pleasure, however. It may be a different kind of self-satisfaction from that of Narcissus, but most of us are probably guilty of some private glow as the effort of planting daffodils is justified in a spectacular spring display. These are certainly flowers to be admired, and shared enjoyment increases the satisfaction immeasurably. Whether it's a roadside clump, a hillside cascade, a whole Field of Hope or just a carefully designed container, daffodils are very inviting flowers. And though they will always haste away sooner than we might wish, the jocund company has usually swelled even more on their return.

The 'Bluebell Fairy' stepping out in 1923, Cicely Mary Barker's
Flower Fairies of the Spring.

Bluebells

A mong the compendious collection of paintings at the National Portrait Gallery in London is a rare depiction of an early European trade agreement. The Somerset House Conference took place in 1604, ending two decades of war between England and Spain and bringing to England an impressive delegation of high-ranking representatives from the Continent. On the English side of the table were the Earls of Dorset, Nottingham, Devonshire and Northampton, all lined up behind Robert Cecil, Viscount Cranborne. On the other side were the Duke of Frías, the Princely Count of Arenberg, the Count of Villamediana, the Senator of Milan, the archducal Audencier of Brussels and the Chief President of the Privy Council under the Archdukes. Though divided by language, religion, loyalties, the English Channel and a very splendid red-patterned tablecloth, the gentlemen on either side are united nevertheless in their fashion sense, for every single one is wearing a large, white ruff. The care with which the artist has recorded the beautifully crimped layers and even concertinas of linen reveals the symbolic status folded into these rather awkward beard-bearing platters. In those days, sense of self evidently rested on a ruff, so firmness was very important. A historic painting of a crucial moment in Anglo-European relations may prompt various questions in twenty-first-century minds, but one of these must surely be, how on earth did they keep those remarkable white collars in order? Those responsible for laundering the linen knew that it all depended on the bluebell.

It may seem strange that such a floppy plant should have been prized as a stiffener, but bluebells were once recognised as a reliable source of starch. Bluebell bulbs contain a sticky substance that can be scraped off and used for creating linen creases as sharp as folded paper. The same viscose paste was already in use before Renaissance ruffs became de rigueur – it proved invaluable for attaching feathers (or 'fletchings') firmly to arrow shafts. This free adhesive was especially efficacious for books and paper, as Richard Mabey found when he tried gluing some sheets together with bluebell slime: 'the paper sheared before the joint gave way'. Such experiments can only be performed with garden bulbs, because the wild bluebell has been a protected species since the Wildlife and Countryside Act came into effect in 1981. Digging up bulbs from woods, lanes or verges is a criminal act. In 1998, an amendment to the act outlawed trade in wild bluebells, too, though a limited number of licensed growers have special permission to supply native bluebells.

Protection of wild bluebells, crucial to the preservation of this beautiful native plant, has meant that what is often sold as a bluebell in markets and garden centres is the Spanish bluebell or its hybridised offspring, long established in the United Kingdom. Spanish bluebells, *Hyacinthoides*

Immaculate ruffs line up as European delegates (left) meet the
British (right) in 1604 at the Somerset House Conference.

hispanica, arrived in these islands in the seventeenth century,
some eight decades after the beruffled gentlemen of the
Somerset House Conference. They flourish in warm, dry
climates, which means that in Britain they are generally to
be found clustering in built-up areas. When the Spanish
flowers cross-fertilise with their British counterparts the
results can be devastating for the natives, because the hybrid
will carry on colonising areas outside urban heat islands,
subduing the indigenous species as they increase. Hence a
certain anxiety in recent years over the survival of English
bluebells, which has prompted lurid springtime headlines
about a creeping invasion of Spanish impostors and foreign
thugs. The taller, more upright stalks and outspread petals
of the Spanish plant, as compared with the demure, bent

heads and trim-waisted bell flowers of the native bluebell, only fuels the suspicion of floral skulduggery. In the past, bluebells have been picked and woven into coronets for crowning busts of Shakespeare on his birthday, and deathday, 23 April. More recently, the bluebell has been proudly hailed on the same day as the flower of Saint George, but even this patriotic claim is a little disconcerted by a reminder that Saint George is also the patron saint of Aragon and Catalonia.

Since the native bluebell thrives in the traditionally cool, damp climate of Britain and Ireland, growing naturally from St Ives to Stornoway, Killarney to Kent, the prospect of steadily rising temperatures poses a further danger. Perhaps this is why bluebells came out at the top in a 2015 poll run by the conservation charity Plantlife to find England's favourite wild flower (the primrose was ranked first in Northern Ireland, Scotland and Wales). There is nothing like the threat of extinction to stir people's feelings. Worries over the possible decline in British bluebells have prompted public flower-spotting campaigns in recent years, such as the Big Bluebell Watch organised by the Woodland Trust in 2017. Enthusiastic observers helped to establish the relative distribution of native and non-native species and provided evidence for assessing trends in bluebell settlement. It turned out that 80 per cent of those reported were the traditional, indigenous plants, with Spanish bluebells still largely clustering in urban areas. Perhaps, most importantly, the call to watch bluebells also sent people out to explore woods and parks and riverbanks across the British Isles, and to see for themselves the annual transformation of winter earth into spring blue.

A sea of bluebells is often the sign of ancient woodland.

This does not necessarily mean a wood filled with veteran trees, but somewhere that has been wooded since before the seventeenth century. Bluebells grow best when left undisturbed, in areas where the fungal mycorrhizae in their roots has gradually become part of a congenial ecosystem. Bluebells open into the broad light that falls through trees before they thicken with summer foliage. Ash and oak woods, especially, leafing late, allow for the light in May. The bulbs lie hidden for most of the year, before sending up their slim, pale emerald leaves and blue, shepherd's crook flower-heads. The perianths hang in rows like translucent blue lampshades, covering their candle-flame stamens. There can be as many as sixteen bells on a single stalk. Glen Finglas in the Great Trossachs Forest or Pengelli Forest in Pembrokeshire are both large areas of ancient woodland and among the best places for seeing bluebells en masse, but even smaller pockets such as the remnants of the great forest on the east banks of the River Foyle in Northern Ireland, or of the Royal Forest at Bernwood in Buckinghamshire, still swirl with bluebells every spring. Close to Great Ayton at the foot of the North York Moors, the Newton woods fill with these flowers in May and, as the slope above washes blue, Roseberry Topping, the distinctive 'Yorkshire Matterhorn', turns into an island crag. Skomer Island, off the west coast of Wales, on the other hand, more or less disappears into sea and sky as the bluebells sweep over it.

It is now relatively easy for many of us to reach a bluebell wood, but in the days when cars were rarer and the rail network far more extensive, some railway companies ran special trains to allow passengers to see the seasonal burst of bluebells along the line. You can still travel through the bluebell woods at Hollycombe in Hampshire, where special bluebell trains run on Sundays in late April and early May. The Bluebell Railway in Sussex runs throughout the summer months, though the Bluebell Special Weeks are from mid April to mid May, when the spring flowers of the South Downs are at their best. Originally, this famous line had the more prosaic name of the Lewes and East Grinstead Railway. When scheduled for closure in 1955, a local campaign was launched to save it and, after a prolonged and high-profile battle, the line finally reopened in 1960 with the rather more picturesque name of the Bluebell Railway. Flushed with success, the campaigners secured an old steam engine, painted it bright blue like Thomas the Tank Engine, and sent it chugging down the line from Bluebell Halt. The Bluebell Railway now has over thirty steam locomotives and attracts around 160,000 passengers a year. Steam trains, once regarded as the hard face of modernity, threatening traditional ways and eroding the distinctive character of different regions, are now seen as the quintessence of an appealingly slower-paced yesteryear, when people had more time to enjoy the changing seasons.

The old local names for bluebells are at home in this half-remembered country. In parts of Ireland they were sometimes known as 'blue rockets', though in Donegal they were also 'crowpickers' or 'bummucks'. Bluebells recalled crow's feet in Lancashire, Wiltshire and Lincolnshire, and cuckoo's boots or stockings across the Midlands, but best of all were 'grammer-greygles', 'granfer-griggles' and 'granfer grigglesticks' of Dorset and Somerset. Bluebells seem such an obvious choice of name for these flowers that it is quite a surprise to learn that they only became widely known as such in the eighteenth century. The *Oxford English*

Dictionary cites 1755 as the earliest recorded use of 'bluebells' in reference to *Hyacinthoides non scripta*, though the word was being applied to harebells two centuries before.

This may account for the different kinds of 'Blue Bell' depicted on pub signs, where some feature a pretty blue flower and others, a cheerful blue church bell. Some cover all possibilities, sporting a bluebell on one side and a blue bell on the other. In the Lincolnshire town of Grantham, now well known as the birthplace of Margaret Thatcher, however, the Blue Bell Inn was only one of numerous colourful hostelries – the Blue Boar, the Blue Bull, the Blue Cow, the Blue Dog, the Blue Horse, the Blue Lion, the Blue Pig, the Blue Ram and the Blue Sheep. Though this fashion long predated the town's most famous daughter, it reflected the heated politics of an earlier era, when a powerful local Whig family, the Manners (whose descend-ants appear again in relation to elderflowers), wielded considerable influence over the electorate through colour-coded drinking. Blue pubs were known to be loyal to the Manners family rather than to the Tory Brownlows of nearby Belton House (perhaps because Colonel John Manners, the famous Marquis of Granby, had been an officer in the Royal Horse Guards – the Blues and Royals). In recent decades, many of Grantham's pubs have been closing down, but if the Sussex railway is anything to go by, it may only be a matter of time before the Blue Bell reopens for a new generation.

Wild bluebells offer refreshment of their own kind. There is nothing quite like the shimmering expanse that turns a woodland floor into an inland sea, where slim trunks stand as if in safe mooring. Bluebells are reminders of the very origins of 'spring', the great gush of life, bubbling up and spreading everywhere as April turns to May. The wetter the winter, the brighter the blue. As a priest as well as a poet, Gerard Manley Hopkins in his 'May Magnificat' presents bluebells as a flower of Mary, celebrated 'through mothering earth', where the 'azuring-over greybell makes / Woods banks and brakes wash wet like lakes'. Though often associated with humility in the Victorian age, for Hopkins bluebells possessed a divine grandeur: 'they have an air of the knights at chess'. He was enraptured by their physical exuberance, their 'shock of wet heads', the 'rub and click' of the stalks, 'the faint honey smell' and the 'sweet gum when you bite them', and gazed in awe at the intricacy of the 'crisped ruffled bells dropping mostly on one side and the gloss these have at their footstalks'. They were flowers full of mystery, baffling 'with their inscape' – the unique identity sometimes revealed in natural phenomena, which was a revelation of the glory of God.

What filled Hopkins with religious awe could also provoke fear in those who saw the flower's startling annual appearance in ethereal blue and smelled the strange scent filling the ancient woodland air. Like bluebells, folk beliefs often survive best in undisturbed landscapes: these plants are traditionally the flowers of fairies (though this is difficult to trace in older writings, because of their elusive name-changing). In pre-Enlightenment British culture, fairy folk were mysterious, unpredictable and rarely benevolent beings. Only the bravest or most foolhardy of mortals would ring bluebells to summon the fairies and if anyone were unlucky enough to hear a bluebell ringing in the woods, it meant that death was coming soon. As English fairies began to lose their more dangerous character in the eighteenth century and to offer inspiration to children's writers

and illustrators of the nineteenth and twentieth, the bluebell's neat petals began to make perfect little caps or skirts.
Cicely Mary Barker, who crafted brilliant observations of
wild flowers, depicted the Bluebell Fairy, tall and slender,
stepping out proudly with one hand on his hip and a staff
of bluebells in the other. *Peter Pan* may owe something to
J. M. Barrie's Scottish childhood, but its immediate inspiration was *Bluebell in Fairyland*, a hit West End musical of
Christmas 1901 about a London flower-seller called Bluebell
and her dream of Fairyland.

The otherworld of fairies often stands for somewhere
beyond ordinary existence, and so for those who feel for
ever exiled from their magical childhood or true home,
bluebells can epitomise the pain of loss. When Anne Brontë,
the youngest of that remarkable literary family, left Haworth
parsonage to become a governess, acute homesickness and
social anxiety made her especially attentive to the 'silent
eloquence' of wild bluebells. Unhappy and far from home,
she longed for 'happy childhood's hours / When bluebells
seemed like fairy gifts' and when she herself 'dwelt with
kindred hearts / That loved and cared for' her. John Clare,
similarly cut off from his home at Helpston and struggling
to cope with the uncongenial new environment of
Northborough, poured his feelings of profound unease into
his poems. In that searing expression of displacement 'The
Flitting', where summer itself has become a stranger, the
first flower in a catalogue of longing for lost friends is
the bluebell. In some folk songs, such as 'The Ash Grove',
bluebells are associated with the intense joy of first love
and remembered, ringing with gladness, from the perspective of lovers who have since parted. The perennial favourite,
'The Blue Bells of Scotland', begins 'Oh where, tell me

where, has your Highland laddie gone'. There are many variations in the lyrics, but rarely more than a single line referring to the flowers – 'He dwells in merry Scotland where blooms the sweet blue bell' – but they somehow match the wistful mood of the song and hence its popular title. For the wild bluebell is a flower of longing and loss. As Ciaran Carson put it most succinctly in his brilliant collection of modern metamorphoses, *Fishing for Amber*, 'The bluebell, flower of mourning, tolls quietly in the dark woods'. Bluebells ring a silent peal for lost childhood, lost home or lost love. Housman's 'Blue remembered hills' in the land of lost content must surely be covered in these flowers.

The bluebell's ancient embodiment of sorrow is recalled in its botanical label: *Hyacinthoides non scripta*. The unusual name derives from the melancholy myth of Apollo's love for a beautiful young man called Hyacinthus. Ovid tells the tale in the *Metamorphoses*, describing how Apollo, god of the sun, music, poetry and healing, played his favourite at discus throwing. Apollo, being immortal, gave the discus an almighty swing, but Zephyrus, the jealous west wind, who was also in love with Hyacinthus, caught it in an extra-strong gust, whirling it away until it crashed into the youth's forehead. The sun god, devastated with grief over the fatal accident, swore to make Hyacinthus immortal by turning him into a delicate blue flower. As the blood drained from Hyacinthus' veins into the earth, a blue plant sprang up, which Apollo inscribed with his grief. English bluebells, also known as wild hyacinths, have no markings on the flowers or leaves and hence their botanical name, '*non scripta*' or 'not inscribed'. The lack of writing on English bluebells not only distinguished them from the Greek species, but also offered a quiet challenge to English poets to give them a literature of their own. John Keats, who loved tales of the ancient world, the spring flowers on Hampstead Heath and beautiful women, imagined the shepherd Endymion asleep in a bluebell bower, being visited by the love-struck moon goddess, who appeared naked except for a spangled, night-sky scarf, like 'the darkest, lushest blue-bell bed'.

The botanical name of the bluebell has been almost as prone to metamorphosis as Hyacinthus. In the early nine-teenth century bluebells were classified in the *Scilla* genus and known as *Scilla nutans*, or nodding squills, until the German plant scientist Christian Garcke reclassified them under the label *Endymion non scriptus*. They were still known at times as *Scilla non scripta* and subsequently as *Hyancinthus non scriptus*. Bluebells are currently called *Hyacinthoides non scripta*, but whether this reinscription will be permanent remains to be seen.

*A gathering of the rapidly expanding daisy family in the Victorian
Ladies' Flower-Garden, by Jane Loudon.*

Daisies

Stepping on the daisies is a traditional sign of spring, though how many daisies are needed varies from place to place. Sometimes it's twelve, sometimes nine, sometimes seven and, in places most eager to see the back of winter, three little daisies are enough. As I spotted three rather puny specimens on New Year's Day 2018, which looked as if they had unwisely put up their white periscopes in very unfriendly waters, their old reputation for seasonal forecasting may be rather misplaced – or another sign of our rural forefathers' determined optimism. It was at least three months before the foolhardy trio were joined by reinforcements. These harbingers of warmer weather have always been linked to the sun. The name, *daisy*, is a version of 'day's eye', the golden sun – or 'eye of day' – that bursts through white clouds. Their white petals open once the light is strong enough; until then, they stand upright, like tiny golf balls on elongated tees.

When a vigorous clump does appear on a golf-green lawn, daisies are rarely welcome. For those whose early summers are devoted to aerating, scarifying, weeding, feeding, mowing and rolling in the pursuit of an immaculate, velveteen expanse, daisies are definitely not on the garden guest list. Apparently inexhaustible, these small, plentiful plants are the bane of the lawn lover's life. It is almost as if the flowers deliberately pull their white sheets over their little, round heads each night to ensure that they wake refreshed and ready to cause havoc. There they are each morning, growing bigger and harder to ignore as the day wears on. Their leaves spread out like miniature squash rackets, as if to offer a playful challenge, but for gardeners whose goal is grass, grass, pure unadulterated grass, they represent the tough, guerrilla cells of an anarchic alliance of weeds. These pest-resistant wild flowers are among the hardest to eradicate: no matter how low the blades are set, no lawnmower can quite destroy their ground-hugging leaves. Daisies can be beheaded and beheaded, but they are still unbowed – a few days in July sees the lawn begin to bristle; a fortnight's holiday ensures complete victory to the yellow-and-white clumps.

The wish to exclude daisies from the golf course is probably reinforced by some of this flower's cultural associations. In the BBC sitcom, *Keeping Up Appearances*, which played on British class-consciousness, the desperate attempts at elegant living by status-aware Hyacinth Bucket (played by Patricia Routledge) were regularly thwarted by the cheerful, beer-swilling, vest-wearing brother-in-law, who is married to her younger sister, Daisy. Among the Bucket sisters, it is the round, cheerful and ever engaging Daisy who cares least about impressing the neighbours. Daisy was just the name for the happy, no-nonsense half of a very appealing but decidedly unstylish couple. The daisy is not only one of the commonest British flowers, but also one of the most popular – except, perhaps, with those obsessed with what other people think. In F. Scott Fitzgerald's classic exploration of the American Dream, on the other hand, a very different kind of Daisy is the unattainable idol driving Jay Gatsby's personal and social ambition.

Daisies have been around much longer than lawns. The so-called 'common' daisy, *Bellis perennis*, is mentioned in the writings of the eminent Roman natural historian, Pliny the Elder. For the English medieval poet Geoffrey Chaucer, the daisy was the flower 'of alle floures', filled with 'vertu and of alle honour', always 'fayr and fresh of hewe'. There was no plant that he loved more and in his poem *The Legend of Good Women* he offers a cameo self-portrait of being up at dawn on a spring morning to see the daisies spreading in the sun. This paean of praise to the daisy opens into a description of falling asleep in the garden and having a marvellous dream, in which the god of love appears with a lady, dressed in green, with a gold net and white coronet on her head, like a daisy. In Chaucer's dream, the flower is the emblem of the perfect woman and devoted wife. As Chaucer was a close friend of John of Gaunt, Duke of Lancaster, and moved among the circles of King Richard II's court, even Hyacinth Bucket might take some consolation from his admiration of the daisy.

Chaucer's daisy praises were influenced in part by the flower's contemporary continental standing. The French name for the daisy, *marguerite*, derives from the Latin, *margarita*, which is related, in turn, to the Greek word for a pearl, *margaron*, because the small white flowers resemble these gems, especially at night when the petals are tucked together, forming tight, white globes. In Renaissance

France, Marguerite was a royal name. The remarkable Queen Marguerite de Navarre, sister of King Francis I and one of the most powerful, brilliant women of the early sixteenth century, took the daisy as her emblem. Beneath her coat of arms, the unmistakable gold-and-white flowers shine out against a background of royal blue, flanked by pink-tipped double daisies, neat as polished buttons – as befitted the royal livery. White flowers were traditional emblems of purity, and the marguerite's special associations with pearls evoked the heavenly 'pearl without price' of Matthew's Gospel. Queen Marguerite's choice also recalled the natural beauty, indomitable energy and tenaciousness of a flower that would be instantly recognisable to all. (A little confusingly, the kind of daisy now usually known in Britain as a marguerite, or Paris daisy, or by botanists as *Argyranthemum frutescens* to reflect the longer, tuftier, silvery slate-grey stems, is an entirely different species, which originated in the Canary Islands.)

Daisies have often suggested constancy: the groundsman's foe is a favourite of the jeweller. On her engagement in 1924, my grandmother, whose name was Margery (a variation on Marguerite), was given a ring with a petite diamond set in a circle of six smaller diamonds, carefully laid out to resemble a daisy. For my grandfather, who was keen on flowers, symbolic meanings and, most of all, my grandmother, this was the best of all possible designs. Years later, when their son was getting married, Margery chose a spray of daisies to complete her wedding outfit, but not the kind that she might find when stepping on grass. It was not long after the Second World War and, with clothes still being strictly rationed, she was sprucing up a pearl-grey suit she had had for some time. Coral, orange or bright pink petals which opened to a span of three inches would make a perfectly contrasting burst of colour, so African daisies – or *Gerberas* – were the order of the day. These flamboyant blooms are widely available in Britain today, but in the late 1940s they were almost unknown and had to be flown in from South Africa. My mother remembers rising at dawn to craft the exotic spray, using the thinnest wire to support the wonderful but worryingly soft-stemmed flowers. Luckily, the *Gerberas* lived through the day and into the family photograph album, though the monochrome film could not do them justice.

Home-grown daisies in spectacular settings can be dazzling, too. Daisies grow along the cliff tops of Dorset, they compete with the well-watered emerald grass from Cork to County Wexford and face out the brisk Atlantic winds along the shores of Skye or Orkney. On the island of Barra, at the southern tip of the Outer Hebrides, daisies glint in the fine grass above the brilliant white sand, making the machair sparkle like a jewelled cloth. But these robust little flowers will also brighten a school field or scrap of inner-city grass. They will pop up in the smallest patches of green, between hospital buildings, by streetlights and bus stops, like little natural blessings. Their proverbial role is to offer a decent covering for graves.

There are taller kinds of native daisy such as the ox-eye, also known as a moon flower or dog daisy, or sometimes, again confusingly, as a marguerite. Vigorous and ubiquitous, like its little look-alike, the ox-eye daisy is welcome in many gardens, while remaining at home in meadows or along the edges of cornfields, hence its botanical name, *Leucanthemum vulgare*, which, unlike that of the 'common' daisy (*Bellis perennis*), emphasises that this white daisy *is* common. Ox-eye is a much more inspiring name, conjuring up a bovine beauty with golden eyes and huge white mascara lashes – or perhaps just a bull's eyeball. As ox-eye daisies can be eaten in salads, vegetarians may prefer to think of them as moon flowers.

The long-standing link between daisies and eyes, so obvious in the popular names and to anyone who looks directly into the flowers or watches them open and close at dawn and dusk, probably inspired the old belief in their power to alleviate ophthalmic complaints. In parts of Scotland, however, children were warned to wash their hands after picking daisies, in case the juice made their eyes sore.

Walter Crane celebrates the comic potential of 'Wide Oxeyes' in his pen-and-verse sequence of pictorial flowers, Flora's Feast, *1889.*

For recent generations of schoolchildren, daisies acquired a new medical resonance through the Heaf test, which until 2005 was routinely delivered to teenagers in the prevention of tuberculosis. Up went the sleeves of the school jumper and out came a special gun to shoot tiny quantities of tuberculin into the exposed arm. A few days later, the appearance of a 'daisy mark' ring of six red pustules meant that the dreaded BCG jab was not required, though it could indicate previous exposure to the disease. In the centuries before, daisies were prescribed for almost everything except tuberculosis. From acne to arthritis, from constipation to kidney infection, from migraine to menstrual problems, from swollen breasts or testicles to stomach ulcers, sprains and sinusitis, daisies were one of the stand-by remedies. Their availability may have been part of the attraction to apothecaries dependent for a living on the medicines they dispensed, and the efficacious 'daisy' was probably an umbrella term for various different plants. The yellow-and-white feverfew, traditionally prescribed for fevers, post-natal problems and freckles, is not unlike the daisy, except for the long, lime-green stems and feathery leaves. Chamomile, another of the apothecary's cure-alls, has similarly daisy-like flowers, branching out on tall, slender stems with green threads of leaves.

The measuring-spoon-shaped leaves of the common daisy were frequently doled out for wounds and bruises, which led to one of its many older names: bruise wort. The famous seventeenth-century herbalist Nicholas Culpeper declared that daisies had the power to heal wounds 'either inward or outward'. Although Culpeper was referring to physical afflictions, the daisy's reputation as a 'wound herb' may well be in the background of the moving elegy

Wordsworth wrote for his brother, John, who was drowned when his ship sank after running into treacherous submerged rocks off the Dorset coast. One of the few sources of comfort Wordsworth found in his bereavement was the 'sweet flower' loved by John, which would 'sleep and wake' upon his brother's grave, as imagined in his poem 'To the Daisy'. The flower's habit of closing its eyes each night and opening them again the next day offered a hopeful image of life after death and a poignant contrast with the eyes of the beloved brother never to be seen again. If Wordsworth was aware of the daisy's power to heal wounds, he was acutely conscious, too, of its associations with accidental death. Robert Burns, the great Scottish poet who died less than a decade before John Wordsworth, had been moved to write about a mountain daisy crushed beneath his plough. One moment it was flourishing, the next it was gone, the unlucky victim of a random accident and an image of the human condition.

When Alfred Tennyson was in Edinburgh, separated from home and family, he, too, was deeply moved by a daisy, which he found pressed in a book. At once, he was transported to the summer he spent in Italy with his new wife, Emily, as they attempted to come to terms with the deep distress of losing their first baby. Tennyson had vivid memories of the sunny orange blossom, the Mediterranean oleander and amaryllis, but the flower that had the greatest impact was a daisy picked for Emily in the mountains above Lake Como. Though the memory was painful, the daisy's unexpected power was a cause for celebration: 'It told of England then to me, / And now it tells of Italy'. Tennyson was capturing the capacity of the ubiquitous little flower to collapse distances and connect hearts.

*Daisy petals measure the warmth of a man's heart in a 1950s
women's magazine.*

The daisy's promise of togetherness is there in the internationally known game of pulling off the white petals one
by one, 'He loves me, he loves me not'. The natural impulse
to connect is embodied in daisy chains, too, as a small slit
in the stalk enables one daisy to link with another and
another until the whole string of daises makes an unbroken
circle. If daisies sometimes provoke irritation, they also
inspire people to join hands and enjoy what they have in
common. The night before I spotted those three forlorn
daisies, I had been celebrating New Year. Once the first
verse and the chorus of 'Auld Lang Syne' have rung around,
the evening's high spirits generally lead to a rather less

clearly articulated rendition of the rest, but as everyone is linking arms and making their contribution to the overall bonhomie, the old song runs on with memories of running about the braes and picking 'the gowans fine'. Daisies – or gowans, as they're known in Scotland – are the flowers of togetherness, whether or not anyone is quite in a fit state to appreciate them.

The anatomy of an elderflower, regarded across nineteenth-century Europe as a medicinal plant, from Franz Eugen Köhler's Medizinal-Pflanzen, 1887.

Elderflowers

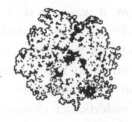

It seemed innocent enough. The day was fine and we had nothing to do. We had had to come in for the last day, but there were no classes, no PE lessons, not even an assembly. I think we had to sign something about returning in September, but otherwise we were free to sit on the field in the sun. Some people wouldn't be back next year – those incredibly grown-up ones, who were leaving school and going to work. It was a day of celebration, at the prospect of the holidays stretching out ahead, and valediction, as those who were leaving made it clear that they were never again going to sit out here, under the beech trees as far from the school buildings as possible. In these memorable circumstances, a bottle of home-made elderflower champagne was just what was needed to create a proper sense of occasion. My friend, Jayne, brought it in specially, and we toasted the day with paper cups. Unfortunately, she left the empty bottle in a carrier bag in the cloakrooms and its next appearance was on the headmaster's desk. What was in the bottle? And was it alcoholic?

These questions did strike us as very funny at the time, which probably answered the second of them. As it was the last day of the summer term, there were not many available sanctions, but when the headmaster very sternly announced that we would not become prefects when we joined the sixth form, it was very difficult to stifle a smile. Traditionally, the elder tree is associated with unpredictable powers of punishment or protection and, in this instance, when its flowers seemed to have brought serious trouble, things all turned out very favourably.

Some makers of elderflower champagne have been less fortunate. In 1993, the Thorncraft Vineyard in Surrey and the British company Allbev Ltd were taken to court for marketing their product as 'elderflower champagne' by the great French champagne company Taittinger, who have always prided themselves on successful cases. Taittinger, who had been growing vines in the French region of Champagne and perfecting production of the world's finest wine since 1734, took a rather dim view of the fizzy non-alcoholic drink that was catching on across the Channel. To use the word *champagne* for such a thing was little short of blasphemy. Since the satisfyingly solid bottles of English 'elderflower champagne' looked very much like the classic French wine, complete with corks and wire, but retailing at a mere £2.50, sales had been rising fast. The High Court judges, agreeing that elderflower cordial was certainly *not* champagne and that such casual borrowing was damaging to *le vrai produit*, granted an injunction. Since then, 'champagne' has been excised from the labels of English wine producers and reserved for the effervescent wine that can only be produced in the French region from which the revered name comes. This does not stop home-made fermentations of British elderflowers, of course, provided that family names remain family secrets. The company who made the contested beverage commercially were not prevented from continuing to market their product, either, but it had to be rebranded. There is in fact a long history of friction between French wine producers and English elder growers, though in the eighteenth century it was elderberry wine that was the problem, because it was often passed off as claret or used to make the port stretch a little further.

Elderflower pressé is now widely regarded as an acceptable alternative to white wine and the sophisticated substitute for lemonade. What was once a homely, old-fashioned sort of drink, associated with grannies and haymaking, makes regular appearances on the cocktail menu. With the increased demand for sophisticated soft drinks, orchards of elder trees have been planted in the Cotswolds, Devon, Dorset and County Longford. The inspiration for these enterprises came from the Vale of Belvoir, on the border between Leicestershire and Lincolnshire, where the first commercially successful elderflower cordial was created in the 1980s. After an exciting and distinguished wartime service in the army, Lord John Manners, brother of the 10th Duke of Rutland, settled down to run a large farm on the family estate in the rolling countryside surrounding Belvoir Castle. In early summer, his wife, Lady Mary Manners, would make her special floral drink for family and friends from the elderflowers growing in the hedges around their home. Gradually more and more people were able to develop a taste for her elderflower cordial, because Lord John started to share a few bottles with local farm shops and then decided to go into wholesale production. Over thirty-five years, elderflower cordial and pressé from the Belvoir Fruit Farms has gone from being a seasonal family favourite to an internationally recognised brand.

Elderflower cordial is only one of many summer recipes reliant on the frothy white sprays that burst from the elder every May. At the height of the flowering period, the trees look as if a parachute regiment has just touched down, their circles of white silk settling into little rucks and wrinkles. Each head consists of clusters of compact, creamy flowers; each flower consists of five circular petals, with five tiny lemon-coloured stamen spades of pollen, all radiating from a pale centre. As the flowers are a mere five millimetres across, each corymb is more of a miniature nebula than a single bright star. Most recipes require armfuls of elder-flowers, so it is just as well that a healthy tree will produce thousands and thousands of these tiny blooms. Elderflowers lose their fresh complexion and honey fragrance all too soon, so unless picked in May or early June, they are already on their way to berrydom. Since overripe flower-heads tend to turn an unappealing brown, which darkens even further if they are harvested and left too long, they have to be quickly transformed into drinks or food, or dried for future use. As the buds open rather randomly, creating necklace-like stems of mixed pearls and tiny white-gold flowers, the optimum moment for gathering the most fully opened but not overblown blossoms is surprisingly short.

All that is required to make elderflower cordial is a quantity of fresh elderflowers, sugar, lemon, water, a large bowl, a strainer and a little patience. Recipes vary: some include wine or cider vinegar, some suggest a different sequence of ingredients, but most recommend a version of shaking the flowers into the bowl, stirring in sugar, lemon

and water, before straining, storing and serving. For alcoholic elderflower drinks, yeast is usually included and a
little more time needed for fermentation. Elderflower pressé,
cordial, wine, or what you will (as long as it's not champagne) goes very well with elderflower sorbet, made from
the opening buds of elderflowers straight from the tree. In
fact, some people claim to enjoy the freshly open flowers
without any added ingredients or elaborate processes,
though they would need to give the heads a good shake
to unhouse unwanted insects. This is an acquired taste,
however, which many choose not to acquire. Most people
prefer their elderflower sieved and sweetened and blended,
whether into gin, jam or jelly, lollies or lemonade.
Elderflower drizzles subtle flavour over summer sponges
and ice cream, or adds an unusual floral fragrance to panna
cotta or syllabub. You can even coat the flower clusters in
batter and fry them into fritters. In the eighteenth century,
foodies disagreed over the relative merits of elderflower and
wine vinegar, but pickled elderflower blossoms were generally seen as a good, economical alternative to capers and
were often made into jams and chutneys. Since elder trees
have always grown like weeds across much of Britain and
northern Europe, their flowers, buds and berries were a
helpful free resource for all and sundry.

Elderflower was the pharmacist's as well as the cook's
friend. A dressing concocted from the ripening buds, soaked
in hot water and stirred into oil, salt and vinegar, sounds
as if it should have been destined for a salad, but was actually prescribed for skin complaints. Elderflower water, or
eau de sureau, as it is known in France (and by English ladies
keen to display their education), was applied as an eye lotion
and skin tonic. It was the go-to potion for blemishes, while

a mix of pressed elderflowers, borax and glycerine worked as an *après soleil* to combat the effects of too much sand, sunshine and seawater. Richard Mabey learned during his research for *Flora Britannica* that young women in the Isle of Man would treat themselves to 'tramman' flower facials to improve their looks. Elderflowers were widely used to lighten skin and clear freckles in the days when the height of beauty was a pale translucent skin, rather than a rich golden tan. The flower-heads were also floated in bath water (usually in muslin bags) to soothe skin irritations and nerves. In centuries past, elderflower water was applied optimistically to those suffering from measles, scarlet fever and anything else that caused rashes. The flowers were fomented in milk for the treatment of piles or mashed with gruel for feverish patients. The monks who settled in the Scilly Isles in 1120 brought the elder trees that gave the principal island its name – Tresco, the place of elder trees. The monastery was both hospital and hospice for the islanders and the seafarers who put into their harbours, so the elders were essential to maintaining medicinal supplies. During the First World War, wounded horses were treated with elderflower ointment, which was made by simmering elderflowers with lard, before cooling and straining. The high level of injury prompted the Blue Cross to put out appeals for huge supplies.

The brief flowering season of the elder tree meant that, until superseded by modern medicaments, elderflowers were routinely dried in large quantities for use during the colder months of the year. They are rapid rotters, so had to be sifted straight from the tree and dried in a warm oven or heated pan. The desiccated flowers could then be stored in airtight jars until an inflammation of the eye or a putrid throat required treatment. They were also salted and pickled and sometimes distilled to extract elderflower oil. Among the most popular of the older remedies was elderflower tea, an international stalwart in the endless battle against colds and flu. It is still brewed up in Bulgaria and produced commercially by a number of herbal tea and health food companies.

This helps to shed light on what might otherwise seem a somewhat surreal watercolour by Arthur Rackham, now part of the V&A collection. The little picture shows an elderly woman in a dark headscarf rising in a tree from the top of a teapot, as sinuous elder leaves gush from the spout.

'In the Midst of the Tree sat a kindly old woman': Arthur Rackham's illustration for 'The Elder-Tree Mother'.

The woman with penetrating eye and pointing finger wears a dress patterned with leaves and clouds of elder-flowers, barely distinguishable from the steam. 'The Elder-Tree Mother' is an illustration for Hans Christian Andersen's tale about the little boy, sent to bed with a cold but then transported through time by a woman who materialises from the steaming tea, which has been brewed by pouring boiling water on to bunches of fresh elderflowers to help his recovery. In this comforting, domesticated, Nordic version of *Aladdin*, the being who appears when the lid comes off the pot is entirely benign and, in any case, her presence is mediated by the boy's kindly grandfather. The Elder Mother begins as a kind of dryad or wood spirit (familiar from Danish folklore as the Hylde Moer), but in Andersen's tale she turns into a figure for Memory and Imagination, revealing the old man's days of youth and the happiness in store for the child. The Elder Mother is both a lifelong loving companion and a beautiful blue-eyed girl, who shows both how natural patterns repeat through the generations and that each has its own special experiences. The sick boy, marooned alone indoors, can still have marvellous adventures, if he learns to look for stories in unlikely places. The metamorphosis of elderflower tea was Hans Christian Andersen's way of re-enchanting the everyday world.

In traditional Scandinavian culture, which left a strong legacy in parts of Britain and Ireland, too, Hylde Moer, the spirit of the elder tree, was treated with respect. Anyone cutting an elder tree was supposed to ask her permission, for fear that she would pursue her own wood, screaming from the fireplace or unsettling cradles and chairs fashioned from it. Elder trees were regarded with reverence as well as fear, their branches brandished against witches and robbers, or brought as blessings to weddings and funerals. In the Scottish Highlands, the juice of elder bark could be squeezed on the eyelids to induce 'the sight' (the special ability to see things which were taking place elsewhere). At Halloween, standing beneath an elder meant that the fairies might appear. In Denmark, it was at midsummer, as the elderflowers faded, that the King of Fairyland was thought to visit those who waited under the elder tree. Hans Christian Andersen was careful not to make his version of the Elder Mother too alarming, for the idea of a powerful and easily infuriated spirit lurking in a teapot might not have helped his audience off to sleep.

Even that inveterate diarist and dendrophile, John Evelyn, could not bring himself to love *Sambucus nigra*. He knew that it was probably the plant most highly prized by contemporary apothecaries, but still refused to give it room on his estate. This may have been partly a consequence of the elder's in-between status – neither a shrub, nor quite a tree – but his powerful distaste smacks of something less rational. The story he tells of the house surrounded by elder trees in which the entire household succumbed to an unspecified disease suggests that he was not entirely immune to the rural superstitions of his age. It may have been the smell that put him off: the pliable, pinnate leaves have a powerful, rancid odour, as different from the sweet musky fragrance of the flowers as can be. It may have had something to do with the tree's physical appearance, for, while a green elder covered in white blossom is the very picture of floral health, an elderly elder can become quite twisted and unbalanced, apparently withering away long before it actually dies. In a small field not far from where I am now, are three antique and oddly angular elder trees, which must once have been part of a hedge. The first in line is the most striking in silhouette, not least because most of it is absent. It has an old, hollow trunk, smoothed as if washed by the sea, which rings out eerily when struck, and though the few small round holes in its side seem made for nesting birds, any that ventured in would be sadly exposed, because the other side has entirely rotted away. Flanking this elderly piece of rooted driftwood are slimmer trunks which, in an odd reversal of the human condition, display strongly lined bark in marked contrast to the bone-like surface of the ancient stump. The younger stems send out leaves in the spring, but there are no flowers yet and even the foliage seems to struggle on the straggling branches. This is a tree that has lost all its pith and promise and yet hangs on to life. It is not difficult to imagine this as the empty home of a weary spirit, yet to be reoccupied – or indeed as the Elder Mother herself.

Christian beliefs often fused with folklore in giving particular plants their cultural charge. The elder was widely regarded as the tree from which Judas hanged himself after the crucifixion, irrespective of its usual height. In some parts of Scotland, its low-growing and twisted habits were attributed to its having provided the wood for the cross (a dubious claim to fame, shared with aspen trees): 'Never straight, and never strong, / Ever bush and never tree / Since our Lord was nailed t'ye'. It would not have been the blossom that inspired these suspicions, but probably the strange brown ear-shaped fungus known, unfortunately, as Jew's ear, *Auricularia auricula-judae*, that grows on elder bark.

Despite such a long history of superstition, the elder tree was known just as widely as a natural toy cupboard. The pretty lace flowers made headbands and necklaces for little girls, wedding dresses or white parasols for dolls. On May Day, elderflowers offered a much less prickly stand-in for hawthorn sprays. Culpeper knew that there was no need to waste time describing it, 'since every boy that plays with a pop-gun will not mistake another tree instead of Elder'. The brittle twigs of the elder tree are full of soft pith, so easy to scoop that popguns, peashooters and pipes have always been made from its branches. This is why it was known as the 'pipe-tree' or 'bore-tree'. The Elder Mother evidently looked kindly on the children who enjoyed her largesse and showered her tiny flowers on the rising generation.

The many faces of the elder haunt Wordsworth's ballad of 'Goody Blake and Harry Gill'. The poem was inspired by a medical treatise by Erasmus Darwin, which dealt with psychological conditions that cause sufferers to confuse illusions with reality, with disturbing physical consequences. Wordsworth was gripped by the report of a Warwickshire farmer, who, after lying in wait on a cold night to apprehend the thief responsible for stealing his wood, was cursed by the culprit with never being warm again and so spent the rest of his life shivering. Goody Blake, a feisty but desperately poor old woman, effectively turns the young farmer Harry Gill's emotional coldness into physical reality ('His teeth they chatter chatter still'). Though based on a true story, the ballad style and Wordsworth's careful choice of firewood charges the poem with suggestions of supernatural power. As Goody fills her apron with twigs, Harry Gill is lurking behind 'a bush of elder'. Cheap, fast-growing and bushy, the elder is a common enough hedging tree, its low twigs and older branches very easy to break off, even for a starving old woman. Anyone aware of the traditional taboo about using elder as firewood, however, would see just how desperate Goody Blake must be. For those who also knew the elder as a tree under the power of a wise old woman, Wordsworth's choice of tree might also expose the young farmer, who had seized what would then be Goody's rightful property, as the real thief in the tale. (It is worth remembering that Harry's attitude was entirely lawful according to contemporary land enclosure acts, which often excluded the poor from ancient common rights of wood gathering.) As Goody Blake has no winter fuel allowance to support her, Wordsworth's ballad distils the perennial problems of provision for vulnerable members of society and the responsibilities of the more fortunate, irrespective of the legal rights and wrongs. 'Goody Blake and Harry Gill' is unsettling because of unexpected revelations, natural rather than supernatural. Wordsworth's elder firewood is a winter version of the elder flowers that bloom in Hans Christian Andersen's tale: both demonstrate the elder's power to reveal extraordinary stories in everyday life. The elder furnishes an international language and a means to articulate what may be difficult to explain in literal terms. Its seeds settle in the minds of those attentive to its strange character and go on growing steadily, surprisingly, like the clouds of flowers in the steam of tea, or the taste for elderflower cordial.

Rosa Damascena Variegata. *Rosier d'York et de Lancastre.*

Damask rose, Rosa damascena, *by the French botanical artist*
Pierre Joseph Redouté. The flowers range from light pink to deeper
red, so it was associated with both York and Lancaster.

Roses

Roses are like days: shortest during the dark months from November to February, when they are often little more than a spiky stalk. Even unpruned wild roses are barely noticeable in a winter hedgerow, unless a rogue briar catches the wind or some unlucky patch of exposed skin. It is when the year begins to stretch and swell into early summer that the astonishing transformation of the garden begins, with bristling yellow or white rosebushes and arching sprays of clustering cream panicles. Wild dog roses, whose blooms are small and pale and layerless, rely on numbers to make their presence known: along roadsides, in railway cuttings, the little fleets of pale pink coracles balance in a sea of green elders, firmly anchored by long, strong stems as the sprays of white spume surge around. Some are white with yellow centres, like delicate poached eggs splashed across a hedgerow. By high summer, garden roses create a prismatic array of iceberg white, old yellow, amber, scarlet, magenta or crimson glory. Walls, sheds and garages disappear under mountainous rambling roses, which hang like suspended avalanches of pink and cream. Roses can shoot up trees to make midsummer fireworks of bright, white-gold star showers, or stay close to the ground, releasing cascades of soft, small spheres over a terrace or rockery. The late flowerers are undeterred by autumn dankness and frost, their cold beauty hung with clear cobwebs, while the bloomless bushes of earlier roses offer round, red hips to ravenous birds.

With so many different varieties of rose, wild and carefully bred, it becomes increasingly difficult to think about 'roses' in general. Yet this is the world's best-known flower and the one most often loaded with symbolism. As a result, it is also the flower most often charged with meaninglessness. Umberto Eco deliberately chose *The Name of the Rose* as the title for his historic novel, because 'the rose is a symbolic figure so rich in meanings that by now it hardly has any meaning left'. I could see he had a point when a casual internet search of 'red rose' produced an opening page dominated by the local bus company, Red Rose Travel. As far as I'm aware, the Aylesbury Vale buses have nothing to do with the England rugby team or with chocolates, or any connection to Lancashire or the Labour Party. Perhaps the company was founded in the 1970s by a fan of Paul McCartney, who regarded the local bus routes as a speedway. Since roses generate fresh meanings as effortlessly as petals, their real importance for humans may in fact lie in their very capacity to sustain a multiplicity of delicate layers while maintaining an instantly recognisable identity. Far from meaning nothing, the rose has often meant everything – especially to those who have gazed intensely into a single flower, rapt by an intoxicating fragrance and by fold after fold of deepening colour.

A flower so gorgeous, various and elusive inevitably piques the collector's urge for order and control. The results are beautiful, perfumed gardens, such as the Europa Rosarium in Sangerhausen, central Germany, where thirteen hectares of multicoloured geometric beds and arches provide the framework for over 8,600 different rose species and cultivars. In Britain, the elegant, brick-walled garden at Castle Howard is home to some 2,000 varieties of modern rose, while the national collection of older roses flourishes in the antique surroundings of Mottisfont Abbey in Hampshire. At St Albans, the Royal National Rose Society has been developing a 'living dictionary' of roses, open to anyone to consult. A database or plant encyclopaedia containing details of the species' appearance, botanical features and history will serve as a reference book for roses, but a *living* dictionary is something else: a multidimensional, full-bodied experience of a huge family of plants, whose special character and subtle variations defy easy classification. The garden in St Albans is unlikely to keep growing, since the National Rose Society had to close for financial reasons in 2017, but a living dictionary could never be complete in any case, any more than the meanings of this flower can be exhausted. Rather than embark on a fruitless quest for the secret rose of ultimate meaning, let me offer a less heroic gathering of buddings, suckers and graftings: an alphabet of roses, beginning with A.

Alice falls down and down into Wonderland and, almost as soon as she is on her feet again, glimpses a tantalising garden through the doorway at the end of a long passage. The rose garden is the loveliest place imaginable, but it is just out of reach – as it has been for generations of children and poets, knights, pilgrims and suitors. The pursuit of the unattainable rose seen in a lovely garden has driven romance and devotional literature since the Middle Ages. In Islamic tradition, too, the nightingale's devotion to the rose encompasses mortal and immortal longings. The intangible rose garden can be an expression of the divine, glimpsed through the dark glass of the material world or of hope for blisses of a more earthly kind. Roses can convey pure beauty and a promise of untold joy – but the fantasy is often fleeting, forever floating out of reach. The iconic image from Sam Mendes' film *American Beauty* was of a beautiful young woman lying in a bath of red rose petals, but everything else in this deeply disturbing film revealed underlying frustrations, deep unhappiness and destructive urges.

For Robert *Burns*, roses were rarely beyond reach for very long. There is no sense of an ongoing quest in his famous declaration: 'My luve is like a red, red rose, / That's newly sprung in June'. The parallel is hardly unexpected, but the startling power of Burns's song is charged by memories of endless lyrical warnings about 'Old Time's' hostility to youth and rosebuds. Almost immediately this rose-red love is likened to something else as well – 'a melodie / That's sweetly played in tune'. Words alone cannot quite capture the full power of love or roses, but the right music might bring us a little closer. At the same time, the perfect union already seems to be slipping apart and, by the final verse, the devoted lover is bidding farewell for a while. Burns knew better than most that the loveliest roses flower and fall: the heartbroken singer on the 'Banks of Doon' tells of a lover who has gone like a rose, leaving her with only the thorn. None of this has deterred numerous song-writers from celebrating love as a rose, mighty or petite, dark, precious or sweet, or from riffing on the flower's power to grow despite snow or to ramble or relive memories or any number of other less typical botanical behaviours. And this is just as well, because anyone who falls in love needs a way to express the overwhelming feeling. A bouquet of red roses or just a single red rose speaks to the moment in a way that nothing else can – in the absence of a florist or an available rose bush, a song featuring roses might just about do instead.

Even those who spend all their time with flowers find that roses have a special hold on the heart. *Constance* Spry, a floral trendsetter in the 1950s, was well aware of the prestige surrounding fashionable modern roses, bred to produce the perfect bloom and last well in flower arrangements. And yet she still retained a soft spot for the old cottage and cabbage roses, with their puffball layers of delicate petals, breathing 'romance, old needlework, old pictures and china, and indescribable scent'. *Rosa alba, Rosa gallica, Rosa centifolia, Rosa damascena, Rosa moschata*, damask, musk, moss: the names roll away like an ancient incantation. Though older roses are prone to pests and disease and only flower once a year, still their special late summer aura and aroma lingers long after they have passed their prime.

The value of *Damask* roses was never primarily dependent on aesthetics or affection. Rosewater, once used as an eye lotion and in cold cream, as well as in eau de toilette, is made from this species, grown mainly in Bulgaria, Russia, the Middle East, India and China. The far more valuable distilled rose oil also requires damask roses, though in France the cabbage rose, *Rosa centifolia* – the rose with a hundred petals – has also been used. Thousands and thousands of rose petals, harvested by hand, are needed to produce rose oil, an essential ingredient of many fine perfumes. So valuable is the oil of damask rose petals, especially in the more refined form of otto – or attar – of roses, that after the First World War the Bulgarian government commandeered the entire national rose crop in order to pay for essential supplies from America.

In the wake of the sudden death of Diana, Princess of Wales, *Elton* John revised his 1970s hit, 'Candle in the Wind', which was originally a tribute to Marilyn Monroe, to '*England's Rose*'. He was bringing together the idea of an 'English rose' (a fair young woman with a fresh, unblemished complexion), the national flower (England has been known as 'the rose' since at least the fifteenth century) and the popularity of 'the people's princess'. His song, already very well known as a reflection on the premature death of a beautiful woman, also drew on the long tradition of linking roses with love, beauty and brevity. All too soon, the budding rose is the rose full blown, though not forgotten.

The flower's habit of rapid disappearance propelled the popularity of repeat-flowering cluster-headed *Floribunda* during the twentieth century. These multiple blooms, achieved by crossing polyantha and tea roses, speak to a desire to amplify and prolong the rose's moment, though transience is as intrinsic to these hybrids as to any other kind of rose. A single floribunda spray can display an entire life cycle, from buds of baby pink to the powder-puff petals of a fully open flower to the forlorn, coffee-stained, screwed-up serviette of a fading rose, and then the bare, spare, spike-collared bristles of a dead head. Unlike the older roses, many floribundas will flower again if the first flush is snipped off soon after the petals fall.

'The rainbow comes and goes, and lovely is the rose'. The loveliness of roses is so well established that these are the chief flowers of choice for anyone whose aim is to challenge convention. The logo of the hard rock band, *Guns N' Roses* – a pair of pistols wrapped in thorny stems like barbed wire with blooming blood red roses – depends on the flower's traditional meaning. The name comes from the lead singer, Axl Rose, its subversive force from the refusal to conform to established codes. The guns and roses are a sign of devotion nevertheless, a badge of allegiance, worn by thousands of fans.

There has always been an appetite for new kinds of rose. During the expansion of global trade in the eighteenth century, the arrival of Bourbon roses and tea roses from China added to the already strained relations between European powers, but the international rivalry led (among other things) to the emergence of a successful *Hybrid tea* rose. The dome-shaped roses from Canton were very different from the familiar forms of puffy cabbage roses or slim-line eglantines, and carried a faint scent more akin to tea than sweet perfume; their soft golden colours and capacity to flower more than once were the real cause of excitement among botany-crazed Europeans. Bourbon roses, a natural hybrid of damask and China roses discovered on a small island in the Indian Ocean, rapidly became established favourites in France. French rose-growing reached a peak of pride in 1867, when the first true hybrid was developed through crossing and recrossing Bourbon and tea roses. At last a beautiful bloom that would flower and flower again was launched under the proud name 'La France'.

France claimed the rose. England claimed the rose. As far as W. B. Yeats was concerned, this flower signified

Ireland. Born to Irish parents, based in London and coming under the general influence of fashionable *fin de siècle* Rosicrucianism and the very particular sway of the beautiful Irish nationalist Maud Gonne, Yeats published stories and poems revolving around the emblem of *The Rose.* His many rose poems were stitched into his nation's 'red rose bordered hem', still trailing a little, but soon to be hitched to hopes of an independent Ireland and personal union with Maud, the ultimate rose.

Yeats's roses tended to be red; the rose of Scotland was white. The Burnet rose, *Rosa spinosissima* (or *pimpinellifolia*) was a *Jacobite* emblem, worn openly as the White Cockade and in secret long after the cause was lost. In the twentieth century, even among those for whom a hereditary monarchy was a matter of abhorrence, this old Scottish emblem stirred deep loyalties. Hugh MacDiarmid, unconcerned with the rose of the world, longed only for 'the little white rose of Scotland, that smells sharp and sweet and breaks the heart'.

Roses bear memories as well as flowers: the '*Kiftsgate*' in our back garden is a reminder of the friend who gave it to us when finally defeated by its extraordinary vigour; of its original home in China; and of the Kiftsgate Court Garden in Gloucestershire, where it was spotted and named by Graham Thomas. Each July, our slender eucalyptus tree, planted by a previous owner, appears to blossom in clouds of cream, because of the rampant 'Kiftsgate' rose clambering about the highest branches.

The *Language of Flowers* or 'Alphabet of Floral Emblems', as pretty Victorian gift books were sometimes called, offered a meaning for almost every flower: 'Angelica . . . inspiration'; 'Bee Orchis . . . industry'; 'Chickweed (mouse-eared)

. . . ingenuous simplicity'. If certain flowers seem a little contradictory – amaryllis apparently signified 'timidity' and 'pride'; golden rod, 'precaution' and 'encouragement' – they were as nothing to the confusions generated by the rose. The list starts uncontentiously by glossing 'Rose' as 'love', but by the thirty-seventh variation, this has come to seem the most baffling flower of all. There are even subtle differences in the same species – a musk rose means 'capricious beauty' but a cluster of musk roses is 'charming'. How easily a well-aimed gift from an eager admirer might misfire if he were not up to speed with this complicated floral vocabulary. Yellow roses, intended perhaps as a compliment to his beloved's blonde hair, or to hint of a golden future, or even a wedding ring, might be translated by a flower phrasebook into 'Jealousy. Decreasing love'. Many a hopeful Victorian nosegay must have fallen very wide of the mark. The enchanting, transporting sight, smell and touch of roses make them the unspoken language of love, but codifying intuition is a perilous business.

So much rose lore turns out to be just a little mistaken, a little misunderstood. The gardens of *Malmaison*, near Paris, where the Empress Josephine famously collected roses as keenly as Napoleon collected countries, were immortalised by the great botanical artist, Pierre-Joseph Redouté. In fact, as Jennifer Potter explains in her excellent book, *The Rose*, Josephine's lovely rose garden grew largely in retrospect, once her memory was being viewed posthumously through rose-tinted lenses. Only two of the paintings in Redouté's renowned collection *Les Roses* are portraits of roses in Josephine's collection.

When Eco chose *The Name of the Rose* as a title because it would 'muddle' his readers by refusing any key to interpretation, he was right. Irrespective of the novel to which it refers, the name of the rose is quite bewildering. Shakespeare's young star-crossed heroine, acutely conscious of her misfortune in possessing a surname as toxic to Romeo's family as his was to her own, declared defiantly: 'A rose by any other name would smell as sweet'. What was important to Juliet was the person, not the label; but the unfolding tragedy shows just how much names matter. In fact the word *rose* is so old and widespread that no other name will really do for these flowers, whose generic name has rambled through Latinate, Germanic and Scandinavian languages, with only a few variations in spelling and pro-nunciation. Although the essential identity of the rose is recognised throughout Europe, this has not prevented thousands of aliases. Rose names refer to the physical char-acteristics of the species, but their extra names may come from breeders or patrons, celebrities or special occasions. Both 'Ena Harkness' and 'Anne Harkness' are named after members of the well-established Harkness family of rose growers; 'Pat Austin' and 'Jayne Austin' after relations of David Austin, who is famous for breeding old-style modern roses. The 'Jane Austen rose', on the other hand, unfolded in soft orange in 2017 to mark two hundred years since the novelist's death. With so many different roses in so many colours and sizes, the room for confusion is consid-erable. 'Eglantyne' is not to be confused with the wild eglantine: it is actually a tea rose named after Eglantyne Jebb, founder of Save the Children.

Although many modern roses take their names from Shakespeare, characters in his plays inevitably refer to the older species. *Oberon*, King of the Fairies knows that the most alluring roses are to be found on 'a bank where the wild thyme grows, / Where oxlips and the nodding violet grows, / Quite over-canopied with luscious woodbine, / With sweet musk roses and with eglantine'. Queen Titania's naturally perfumed bed, with its thicket of protective briars, prompts lyrical longings in her temporarily estranged husband: roses tend to come to minds in the absence of partners, as in Robert Herrick's musings upon 'a red-Rose peeping through a white'.

A warm, July afternoon in a rose garden is heavy with *perfume*, but nothing quite compares to the pleasure of brushing against a fresh rose at dawn – as the air is charged with the unexpected rush of scent. Even their leaves smell faintly of spices when rubbed. Fresh rose petals can be crushed and boiled to make a fragrant jam, used to perfume butter for rose-petal sandwiches, or frozen for floating in gin and tonic. *Potpourri*, created from sun-dried roses blended with salt, lavender, mint and bergamot, preserves the scents of summer and makes bedrooms and bathrooms redolent of a National Trust shop.

Queen Elizabeth I, who understood the importance of image in maintaining power, often chose to wear necklaces of jewelled roses when sitting for a new portrait. She followed the lead of her grandfather, Henry VII, who had bolstered his claim on the throne by deploying the new Tudor rose. Elizabeth surrounded herself with rose iconography as she promoted the Tudor dynasty and attempted to divert adoration from the Holy Mother of the Catholic Church to the virgin Queen of the new Protestant nation. In her later years, she became increasingly inspired by poems in praise of England's 'brave eglantine'. On her sixty-second birthday, George Peele urged everyone to 'Wear eglantine / And wreaths of roses red and white' in celebration. When Oberon described the fairy queen sleeping on a bank of sweet musk roses and eglantine, he was paying a compliment to Elizabeth as well as to Titania. The prickly eglantine also warned potential suitors to approach at their peril: this pure, natural, if now rather mature, rose was definitely not the one to pick. Modern Queen Elizabeth roses – pink, healthy and straight in stature – are often used, perhaps irreverently, for creating thick and prickly hedges. The perennial royal rose connection is strengthened by thousands of heavily scented roses in Queen's Garden in Regent's Park, originally planted

Queen Elizabeth I, adorned with jewels and a pink rose on her lace collar, in the Ditchley portrait by Marcus Gheeraerts the Younger.

under the stern eye of Queen Mary, patron of the national Rose Society. There are numerous monarchically named roses, though the rose known as 'La Reine Victoria' is a French Bourbon rose, as is 'Prince Charles'.

The *Rose Society* of the United Kingdom was founded in 1876 and survived until 2017, when it was forced into administration. The official aim of the original foundation was 'to encourage, improve and extend the cultivation of the rose'; the unofficial aim was to compete with France, where concerted crossings had resulted in those very desirable hybrid tea roses. Annual prizes for innovation spurred British growers to come up with their own modern champions, and soon became a regular source of horticultural pride and perspiration. Roses are judged according to their growth, beauty, fragrance, flowering habit, freedom from disease and crucially, 'general effect'. The experts know that each of the rose's many attributes deserves special praise, but there's also the overall wholeness of the perfect rose. The desire to produce the rose of all rosiness means that prize-winning blooms are reared in hotbeds of anxiety. Modern rose growers may not have the threat of execution hanging over them like the hapless gardeners in Wonderland, so busy painting a white rose red, but the determination to grow the perfect bloom in just the right place, or at just the right time for the annual competition can seem almost as stressful. Whether it's a professional breeder preparing for a national contest, or a private gardener determined to scoop the cups at the village horticultural show, the pressure is intense. The quest to identify or develop rain-resistant strains is especially determined among British rose growers. Throughout June and July, roses are as prone to being ruined by rain as the village fetes, music festivals, garden parties, cricket matches, weddings

'The roses . . . were white, but there were three gardeners at it,
busily painting them red', Alice in Wonderland.

or fun-runs that continue to be organised in the annual
national triumph of optimism over experience. For proud
gardeners, who have been nurturing their prize specimens,
picking off aphids and protecting against black spot, rust or
mildew, a sudden cloudburst can seem a disaster of biblical
proportions – as the stems buckle in the blast and the huge,
perfect heads collapse in a cloud of cascading petals. Even
an unexpected heatwave can cause havoc. Petals, floating
free, drift gently in the air for a second or two, but land all
too soon. The colour drains and the form dissolves and
nothing remains but a white rash on the lawn.

The least weather-dependent garden of all is the *Secret
Garden*. Here roses live in the mind, in an unchanging

haven of warmth, colour and heavenly scent. This is a garden remembered from earlier, sunnier days, and it holds the hope of recovering wholeness, of happiness entire and of itself. T. S. Eliot reflected wistfully after visiting old roses in a Cotswold garden on an incomparable summer day, that secret rose gardens can remind us of 'the passage which we did not take / Towards the door we never opened', as well as of the choices we made and came to wish we hadn't. Here the rose garden is an image of an earlier self, opening unexpectedly on the present and bringing home a sense of permanent exile from paradise. Rose gardens can mean regret, or they can mean joy realised in the company of another and hidden from the rest of the world. Secret rose gardens can be promises too, signalled by the faint fragrance of roses as yet unseen. In Frances Hodgson Burnett's well-known book, *The Secret Garden*, the lonely heroine, fascinated by the housemaid's story of a locked garden somewhere in the grounds, is determined to find it. When she finally succeeds in unlocking the hidden door, she finds a garden overrun with climbing roses and rose clumps, though as it is winter there are no leaves or flowers or colour, only a 'sort of hazy mantle spreading over everything'. Gradually, the garden begins to come back to life, 'And the roses – the roses!' are at the heart of an all-encompassing transformation of everything and everyone in the book.

We are told that in the Garden of Eden roses grew 'without *Thorn*'. To regain paradise, then, it might be worth seeking out roses such as 'Tranquillity' or 'Sleeping Beauty', which have been bred to have very few thorns. Roses produced for the cut-flower market are often deprickled, too, to save unsuspecting fingers. As a general rule, the wilder the rose, the thornier, though I have often found my hands covered in scratches and the dark pinheads of thorn tips after some impulsive, gloveless pruning. One afternoon in the garden ended with a poisoned finger and a trip to A&E. The only thing to be said for the incident was that it gave my son a chance to try lancing the infection with a home-boiled safety-pin before we went to hospital, and strengthened his ambition to become a doctor.

If roses sometimes mean pain, they also mean peace. The first meeting of the newly formed *United Nations* took place at the end of the Second World War. When the delegates arrived from around the globe, each was given the new 'Peace' rose, launched in the United States in April, on the day when Berlin fell to the Allies. The huge hybrid tea had been bred in France in the 1930s by Francis Meilland; as war loomed, he had the foresight to send cuttings abroad for safe-keeping. The fat buds of this lovely rose begin to crack with a thin, gilt stripe and gradually reveal a fresh blood-coloured streak, before opening into glowing heads of enormous proportion, golden with pink waves like the dawn-drawn clouds of a summer day. These are the roses of a new morning, but they keep on opening and come back again year after year. Of the thousands of roses now being grown, the enduring favourite is 'Peace'.

Botticelli saw *Venus* emerging from the sea in a huge shell, naked except for a shower of roses scattering from above. The rose is also a traditional symbol of the *Virgin* Mary, an embodiment of pure love in the Catholic Church. Though their followers idealise love of different kinds, the devotion inspired by these divine female figures is intense. The roses proffered on St *Valentine's* Day can also convey different kinds of love, as I discovered when working at an American college one winter. The students devised a special litmus test for

14 February, when each table setting was adorned with paper roses from supposedly secret admirers, who chose white, pink or scarlet petals to show whether their love was Platonic or more passionate. The evening saw lots of people happily brandishing scarlet bouquets and lots more trying to conceal one or two, kindly meant white flowers.

Devotion can have its own destructive consequences. The *Wars of the Roses* tore England apart during the fifteenth century, as the blood-red rose of the Lancastrians was pitted against the white rose of York. 'X' meant crossed swords and crossbows, and now marks the fields where the battles raged, at Barnet and Northampton, Ferrybridge and Towton, St Albans, Hexham, Tewkesbury, Wakefield, Worksop and Mortimer's Cross. The long war finally came to an end in 1485, when Henry Tudor defeated Richard III at the Battle of Bosworth Field and then united the red and white roses in his special Tudor rose. When Richard's body was discovered many centuries later under a car park in nearby Leicester, he was given a royal reburial in the cathedral, where his connection to the House of *York* was recognised in linen cloth embroidered with the white *Yorkshire* rose. The Yorkshire County Cricket team still sports the badge of the white rose. Pub signs, bunting, cycling clubs, scaffolding companies, and animal rescue centres across God's own county proudly feature the Yorkshire rose, though the gardens at Castle Howard do admit red roses as well as white.

I was introduced to '*Zéphirine Drouhin*', an old, rambling rose with the smoothest of stems, by my old neighbour, who loved all roses, but this one in particular, which grew on the warm wall around her cottage door. I planted one for her in our new home after she died, and every year the deep pink roses bloom in quiet commemoration.

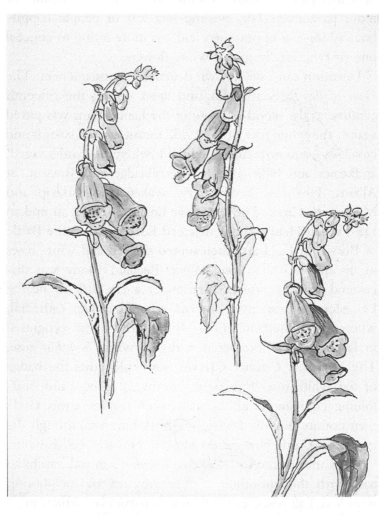

Beatrix Potter's studies of sinuous foxgloves (and a small abandoned sketch of a robin): preparation for The Tale of Jemima Puddle-duck.

Foxgloves

When Jemima Puddle-duck waddles out of the farm-yard in search of some peace and privacy for hatching her eggs, she lifts off and flies over a distant wood, landing 'rather heavily' amidst the trees. As she looks for a secluded haven for nest-making, she spots an elegant gentleman, sitting on a tree stump surrounded by foxgloves. Jemima, herself, bonneted and enshawled, is dwarfed by a pair of these towering flowers, hung with tiers of pink petals like joints of ham. A more observant puddle-duck might have realised that the long-fingered blooms were pointing to the true identity of the gentleman with the sharp ears. Even when she follows him to his 'dismal-looking' summer residence in the depths of the wood and cannot fail to notice the bushy, tawny, white-tipped tail protruding from under his tweed jacket, Jemima still does not grasp that the foxglove grove is home to a fox. What follows is comic but cautionary, as the rather gullible duck has a very narrow escape and the cunning gentleman with the sandy-coloured whiskers gets his comeuppance. Beatrix Potter was warning her young readers that not all well-mannered gents turn out to be quite as nice as they may seem.

Since foxgloves grow abundantly in the Lake District in early summer, Beatrix Potter would not have had to walk far from her new home in Sawrey to see them flourishing in the thin soil beside a beck, breaking free above fresh bracken, or licking the sky behind a dry-stone wall like strange purple flames. They are still the easiest of wild flowers to spot, scattered along farm tracks, balancing on steep scrubby slopes, massing in a grassy dip or woodland clearing. Foxes often build dens in rough, unkempt copses, hidden among tangles of brambles, old tree roots and fallen branches. These are sites where foxgloves thrive, too, tall enough to stand their ground even in the face of tough competition from brambles and nettles. Often they grow in the disturbed ground surrounding rabbit holes, serving as restaurant signs for foxes. Seamus Heaney remembered the tall foxgloves around the old wells on the farm where he grew up in Derry and the rat that ran from them, scaring the boy who stared into the dark water, as it 'slapped across his reflection'.

Foxgloves have always been known as flowers to approach with care: the roots, leaves and flowers are all poisonous and, though rarely fatal, would make an unlucky plant forager very sick indeed. Children, rarely averse to risk, will pick the little purple bells to make detachable claws for their hands, unknowingly affirming the meaning of the flower's botanical name – *Digitalis*, or 'finger-like'. 'Foxglove' suggests that foxes would pop the bells over their paws, perhaps when engaged in nocturnal robberies, but they are also known in various counties as 'granny's gloves', 'fairy gloves' and 'thimble flowers'. In Scotland, these flowers were thought to be favoured by witches and hence their old name in the Borders: 'witches' thimbles'. The slight droop at the tip of many foxgloves may also have brought the traditional pointed hats of witches and wizards to mind. The Scottish writer, Katharine Stewart, who worked hard to create a garden further north near Loch Ness, remembers being stopped in her tracks by the sight of a tall white foxglove growing up from the nearby heather. Although it probably seeded itself from her garden, she recognised it as 'a signal' and immediately set off across the hillside 'to the remains of what was known as the witch's house'. There was nothing very frightening about the memory of this 'witch', who was known to possess 'much goodness in her heart and patience with the young', except that she lived alone in a tiny, tattered thatched cottage, with a leaking roof and little means of keeping warm.

Walter Scott was well aware of the flower's more dangerous reputation and included 'Nightshade and Foxglove side by side, / Emblems of Punishment and pride' in his hugely popular poem, *The Lady of the Lake*. Scott's foxglove keeps dangerous company, since the nightshade is notorious for its deadliness and capacity to mete out capital 'punishment' to those who rashly consume its berries. *The Lady of the Lake* put Loch Katrine at the top of the trail for nineteenth-century tourists, who flocked to the Trossachs, where foxgloves can still be seen growing wild among the steeply sloping moors and open woodland. Why the foxglove was emblematic of pride is immediately evident in its lofty demeanour and capacity to rise far above other wild flowers in the abundant undergrowth of July. A vigorous spike will grow to four or five feet in height, displaying as many as forty magnificent bells. The foxglove shows off its numerous orifices with reckless defiance: inside the wide open mouths those long pale throats, speckled with dark crimson blotches and white ulcers, look decidedly putrid. These flowers are also known as deadmen's bells.

It may have been the unhealthy-looking lips of the foxglove that prompted herbalists to recommend the leaves as a treatment for both ulcers and scrofula, which could cause very distressing and unsightly growths. In Italy, where wild foxgloves thrive from the Dolomites in the north to Calabria in the south (though seemingly not along the south-eastern coast of the Adriatic), the bruised leaves were used for cleaning and binding wounds. When the Scottish doctor and antiquarian, Martin Martin, travelled through the Western Isles in the early eighteenth century, he reported that people in Skye would plaster warm foxgloves over the site of post-febrile aches and pains. Elsewhere, foxglove leaves were boiled in hot water or wine as a cure for breath-lessness, though too much of this brew might well have made the patient's breath cease altogether. In Derbyshire, foxglove tea was taken recreationally, according to a 'Naturalist's Report' in the *Time's Telescope* in 1822, which described its popularity among the local women 'of the poorer class'. With no expense and very little effort, foxglove leaves apparently produced 'great exhilaration of the spirits, and other singular effects on the system'. Jemima Puddle-duck's gentleman hardly needed to resort to offering foxglove tea to such a willing house guest, but this brew may some-times have served as an early version of the spiked drink.

The foxglove's powerful properties have been observed for centuries, but it was not until 1785 that the real thera-peutic benefits began to be understood properly, after the botanist and doctor William Withering published his pioneering *An Account of the Foxglove, and Some of its Medical Uses: With Practical Remarks on DROPSY and other Diseases.* This was the first serious publication on the medicinal power of digitalis. Withering's work was based on ten years' medical practice of treating congestive heart failure, but his interest in digitalis went back to the herbal remedies in use during his boyhood in rural Shropshire. In his experimental research on an effective treatment for congestive heart disease – or dropsy – Withering recalled the old woman who had prescribed handfuls of herbs for complaints of this kind, sometimes with startlingly positive results. As he worked through the herbal cornucopia of possibility, Withering discovered that the active agent in the old remedy must have been foxglove, which he realised had unparalleled powers over the pace of the heartbeat. Although many of his trials proved inconclusive, Withering persevered and his published *Account* was rapidly recognised as a major scientific breakthrough, causing ripples in medical circles throughout Europe and America. As news of the wonder-drug spread, people were quick to seize a retail opportunity. Covent Garden flower stalls started selling foxgloves to ordinary people, who would boil up handfuls of the leaves to cure dropsy, and sometimes survived.

Withering recommended tiny quantities of dried foxglove leaves, boiled with water and cinnamon, macerated, filtered and carefully dispensed, to be given twice a day to those suffering from dropsical complaints. Other doctors began to experiment, sharing their practice and comparing cases. Mr Brown of Muscovy Court, who was suffering from a 'violent palpitation of the heart', was given a teaspoon of digitalis tincture every four hours until his heart rate began to change, at which point the dose increased to three ounces. A young lady (unnamed) with shooting pains and night sweats was given three grains of digitalis powder every day for three weeks. According to Robert Thornton's report, both patients made a full recovery. Sometimes the flowers were bruised and mixed with lard, sometimes simmered in water; usually it was the leaves that were dried, before being boiled in water or alcohol and dispensed. Concerns about the drug's potential dangers, about which parts of the plant were safe to use, about increasing or decreasing the dosage and about how far the pulse should be allowed to rise or fall, spurred quill pens into frequent correspondence. John Keats, training for his medical examinations at Guy's Hospital, made careful notes about digitalis while writing sonnets featuring foxglove bells in his spare time. By 1810, nine distinct complaints had been identified for treatment by digitalis in various forms – inflammatory diseases; active haemorrhages and phthisis; anasarcous and dropsical effusions; heart palpitations and aneurism; hydrocephalus; mania; spasmodic asthma; scrophulous tumours; and epilepsy. This may seem a different world from that of the modern high-tech hospital, but digitalis is still in use in the form of digoxin for atrial defibrillation and heart failure – though doctors no longer have to harvest foxglove leaves for their pharmaceutical supplies.

The foxglove used for the modern production of digoxin is *Digitalis Ianata*, or Grecian foxglove, a splendid, substantial spire of mushroom-coloured bells, which thrives in its native fields of central Europe and across much of America. Unlike their purple relations, these foxglove bells have protruding lips which are neither blotched nor spotted, but run with dark red veins. There are, in fact, quite a range of foxglove species. When John Clare was studying the habits of bees in the 1820s, he was struck by the 'very curious bee' that buzzed about 'the Iron brown fox glove'. The southern European rusty foxglove, *Digitalis ferruginea,* was already common in English gardens in the eighteenth century, and Clare was noting its colour because he was trying to assess whether certain insects were drawn to particular species of flower. Foxgloves come in different colours, from rusty brown or apricot bells to the pale yellow of the Siberian foxglove, *Digitalis Sibirica.* The smaller yellow foxglove, *Digitalis lutea,* with its sparser and less spreading bells, grows well in southern Europe and north-west Africa, but will also survive lower temperatures than its Siberian counterpart. All of these graceful, strikingly coloured flowers have found a place over the years in the borders and shady corners of British gardens.

The tall, tapering flower fingers of the common foxglove, visible in the wild above rampant summer undergrowth or behind the lower-growing plants in a shady flowerbed, are the most successful in beckoning bees. No colour is more attractive to a roving bee than purple, so these spires of appealing mauve bells are calling the bees to their meals. The spreading mouth of the foxglove petal will accommo-date even the fattest bumblebee: its flat lower lip is a natural heliport, allowing safe landing right outside the bar. The bee bustles into the narrower passage to get its fill, unde-terred by the hairs that keep smaller insects at bay. Satisfied and coated in golden pollen, the bee backs out, ready for take-off and another foxglove snack in exchange for its load of gold. Foxgloves may be plants to handle with care, but their contribution to the ecosystem has always been – and still is – immense. By sustaining bees, they indirectly sustain us all.

'Burt's Seed for Quality': foxglove seeds from the 1910s, marketed by New York seed supplier William D. Burt.

A century after Clare was making his notes on the natural history of Helpston, the 'wild and woodland gardening' movement was sweeping the United Kingdom and putting foxgloves at the height of horticultural fashion. Garden designers recommended colonies of foxgloves for planting 'along the fringe of a wooded boundary', where they stood out in bold outlines against the darker shade of the trees. Not only were foxgloves aesthetically appealing, but also their self-seeding habits meant that, without excessive outlay, even a very lengthy fringe might soon be thick with tall spikes. The Chelsea Flower Show of 2018 demonstrated the perennial appeal of these living architectural forms, with displays of pale foxgloves silhouetted beside polished concrete steps and columns to create a calm, minimalist haven. Foxgloves are very beautiful plants and, whether growing in the quiet order of a garden or in shady summer woods, they always seem to be rising above the tangles of life and pointing towards the heavens. The white common foxglove, *Digitalis purpurea alba*, was much in demand in the 1950s, after Vita Sackville-West and Harold Nicolson created their famous White Garden at Sissinghurst Castle. Within the old stone arches and new yew hedges, no flower was to be admitted unless it was white. It took some time for the plans to take shape fully, but from the beginning colour was subdued and form accentuated to create a cool haven set apart from the hot hues of the summer garden. Vita even imagined a 'great ghostly barn owl' sweeping across their pale garden at twilight. While birdwatchers may baulk at the idea of colour co-ordinated flora and fauna, the luminous pallor of a white garden does seem a fitting setting for nocturnal flight. A barn owl moves like breath across moonlit grass, a sudden, soundless presence that is gone almost as soon as it appears. Whether it would bother with an enclosed garden, when there were open fields full of mice all around, is another matter – this was an aesthetic rather than eco-logical vision. And here a white foxglove could come into its own, as finely shaped as the sculpted pinnacles of a medi-eval cathedral, a silent spire of unrung bells.

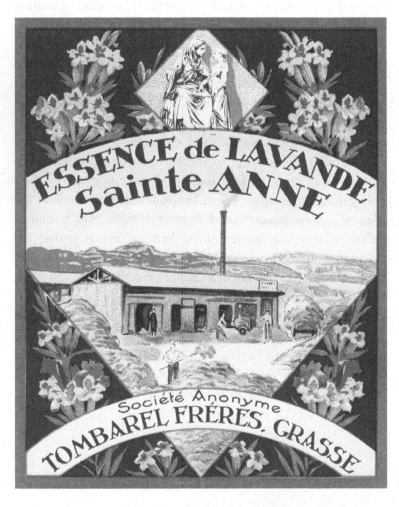

Lavender essence label from the Tombarel brothers' perfumery in Grasse, in the south of France.

Lavender

It was a hot, early summer morning in the heart of England. A narrow lane, fringed with long grass and cow parsley, twisted and turned through bristling hedgerows, beside fields of butter-coloured barley and grass pastures that fell away to invisible streams. Scattered, golden stone cottages, with gardens spilling out in clouds of pink and white, seemed outgrowths as well rooted as the clusters of ashes and oaks. The road dipped round an unassuming bend and, suddenly, things were not as they had been. The hill directly ahead was blue. Deep blue. Sapphire blue. A blue that seemed to drain the summer sky of its delicate complexion. A huge field of lavender filled the horizon, looking as if an old giant had turned over in bed, pulling a cobalt counterpane over his shoulder to shut out the morning light. Lavender plants, native to the Mediterranean, thrive on dry weather and hot sun, which is why they are not always at their best at the beginning of a British summer. In some years, though, the damp mists of spring make their final exit by May, and then the trees, rising from the hedges in fresh spring green, stand out in high definition to usher in a warm, cloudless June. At times such as this, lavender buds can burst open in bold blue, lilac or mauve to seize the limelight from all the unready cereal crops around, as England unexpectedly begins to assume the hot, bright colours of Provence.

For Gertrude Jekyll – the brilliant turn-of-the-century horti-colourist who turned gardens into works of art, learning from Turner's paintings to create her own vivid, living drifts – this flower was a favourite in the planting palette. A perfect companion for paler flowers and straggling boughs of grey-green rosemary, an amiable neighbour to spreading blue catmint, silver-grey lamb's-ears stachys or pastel and white night-scented stocks, lavender quickly turns into bristling mounds of imperial purple or royal blue. If glimpsed through a bed of bright orange nasturtiums, tawny snapdragons or yellow roses, lavender grows bluer than ever. Jekyll would plant cascades of soft, colourful flowers and complementary borders at her home in Munstead Wood in Surrey, not to mention the four hundred or so other gardens she was invited to transform. 'Munstead lavender', a dwarf variety of the slim-stemmed, spiky species *Lavandula angustifolia*, is still much sought after for its strong scent, classic lavender-blue colour and ebullient habits of growth: this is a perennial best-smeller among decorative hedges. In Jekyll's hands, lavender softened the borders between lawns and paved paths, stone steps and lily pools, terraces and rose beds and, being an evergreen, between summer colour and winter drabness. It flourished equally well on the border between use and beauty, a special zone that Jekyll understood well, 'where pleasure garden meets working garden'. Lavender remained a standard of cottage and kitchen gardens, even as it began to occupy a front row seat in the herbaceous borders of country houses. There is nothing too difficult about growing lavender, provided it has full sun and regular pruning (lavender plants can become rather wooden-leggy, if left to age untended).

Long before it became valued for its visual appeal, lavender was prized for its therapeutic powers. The medics of ancient Rome mention lavender for stings and bites, stomach disorders and chest complaints. From the Middle Ages it was regarded as a wonder-plant, widely credited with saving, or at least soothing, troubled bodies and minds. The remarkable, multi-talented German Abbess Hildegard of Bingen relied on lavender to deter lice and encourage sound slumbers. The oil distilled from lavender flower-heads, which could be mixed into medicinal cordials, was used to treat a steadily increasing array of ailments. The 'lavender drops' offered to Marianne Dashwood in *Sense and Sensibility*, when she is mortified by Willoughby's rejection, were probably a tincture of lavender oil. A year before the novel appeared, the contemporary medical botanist Robert Thornton included a recipe for 'Spirit of Lavender' in *A New Family Herbal*, which required pounds and pounds of lavender flower-heads, blended with rosemary, cinnamon, nutmeg, sugar and red sanders (or sandalwood *Pterocarpus santalinus*), which made the tincture more palatable by turning it a pretty red. It was highly recommended for those suffering from hysteria, lowness and any other nervous afflictions. Thornton's remedy was a simplified version of an old favourite commonly known as 'Palsy Drops' or 'Red Hartshorn' – an extravagant concoction, requiring up to thirty herbs, flowers and spices, blended with brandy. Over the centuries lavender, in a variety of forms, has been prescribed for everything from indigestion to epilepsy, toothache, hysteria, head lice, depression, cramps, palpitations, oedema, vertigo, migraines, fainting fits and snakebites. It was also the best thing for bathing sore feet, which may be why Aunt Jobiska made the Pobble who has no toes drink 'lavender water tinged with pink' in Edward Lear's memorable poem.

After the French scientist René-Maurice Gattefossé discovered that lavender oil was analgesic and efficacious in the treatment of burns, it was in great demand during the First World War, when gassed and blasted soldiers were often left suffering the aftermath of battle for many weeks. Although its antibacterial properties are now known to be less powerful than was once assumed, the refreshing scent of lavender oil may have contributed psychologically to the healing process. That lavender has ancient connections with cleansing is evident in its name, which is probably derived from medieval Italian *lavanda*, or 'washing' (though there is a theory that it comes from the Latin *livere*, meaning 'to be blueish'). This may have enhanced its appeal to those scarred by the trenches, not least because of its reputation as a treatment for lice. The soothing, intensely aromatic plant, traditionally prescribed for calming nerves, lifting spirits and helping insomniacs, was a reassuring resource for the wounded, the shell-shocked and those battling to aid their recovery.

Lavender had been an important crop in France for many centuries before its flowers were needed for the field hospitals of the First World War. French perfumery had relied on the purple fields of Provence since the first manufactured fragrances began to waft around Grasse in the fourteenth century, though it was only in the nineteenth that lavender began to dominate the industry. Lavender-scented eau de cologne is still regarded in France as part of a man's toilette rather than being a woman's fragrance. The lavender plants now known in Britain as 'French lavender' – *Lavandula stoechas*, with bright tufty heads like bursting purple pineapples, and *Lavandula dentata*, distinguished by slim, green, scallop-edged leaves – are both islanders by origin, though their first homes in the Îles de Hyères, Madeira and the Canaries enjoyed rather gentler winters than England, where they arrived in the mid sixteenth century. Though now well settled, they still remain a little less hardy than what is generally known as English lavender, the spikier *Lavandula angustifolia*, which was probably ferried to these shores in a Roman galley some centuries before.

Although lavender reached its peak as a popular perfume in the Victorian era, it had long been in use for freshening homes and their inhabitants. The Elizabethan poet and farmer, Thomas Tusser, regarded lavender as a herb for strewing on floors or putting in a pot on the windowsill, along with various other fragrant herbs and flowers. In medieval literature, 'lavender' was a name sometimes given to washerwomen and, though referring to their occupation, helped strengthen the association between lavender plants and laundry. For centuries afterwards, linen kept its cleanly laundered smell in the summer months with the help of lavender sprigs and, as the flowers retain a strong scent long after cutting, the dampness of autumn and winter could be disguised by sprinkling dried lavender lanterns over beds and rugs. (Lanterns are made by binding stalks, folding in the head and weaving ribbons in and out.) Pretty little lace lavender bags suspended in wardrobes made clothes seem similarly fresh and kept the moths at bay. Modern lavender-scented drawer liners and aromatic furniture creams and polishes are heir to a very long domestic tradition.

During outbreaks of the plague, lavender was proffered to ward off the illness and disguise the smell of those unfortunate people who hadn't quite managed to do so. In less critical times, it offered inspiration for combatting perspiration: by the nineteenth century, women were dipping themselves in lavender-scented baths and washing with lavender soap. Meanwhile their gentlemen friends were slicking back their hair with bear grease and lavender oil, patting their freshly shaved faces with lavender water or slicing dried lavender to smoke in their pipes. For non-smokers, a smouldering bundle of desiccated stalks would fill the air with a faintly intoxicating aroma. The insatiable demand for all these perfumed products made growing lavender highly lucrative.

There was a time when lavender fields filled acres of Britain. Two hundred years ago, July in Croydon would have been heavy with the scent of lavender, as entire families harvested armfuls of this valuable plant. It was hot, hard and often hazardous work, as the strong scent and sweet nectar of lavender is a magnet for bees. As a long-standing cottage industry which had grown up in the congenial environment of the North Downs, it was also a defining feature of local culture. Mitcham was more or less synonymous with lavender, because of the numerous farms that had made it a centre of production since the sixteenth century. The first English toiletries company was founded in the Surrey town in 1749 by the medicinal gardening partners and entrepreneurs, Ephraim Potter and William Moore, and continued to boom throughout the century following, when the fashion for lavender reached its peak.

'Genuine Mitcham Lavender Water' produced and promoted by the Surrey lavender and peppermint grower John Jakson in the 1890s.

Yardley, the English perfume company, caught the rising tide perfectly when they launched their famous English Lavender brand in 1873. Although twentieth-century tastes largely turned to other fragrances, English Lavender retained an increasingly quaint and olde-worldy appeal for many American consumers.

By the 1920s and 1930s, as demand for houses and shops outpaced the market for lavender, the English industry was in serious decline. The old Surrey farms were grubbed up and gradually absorbed into greater London. Norfolk, a little further from the swelling capital city, just managed to keep its lavender fields alive, though by the 1980s the Norfolk Lavender Company at Heacham was the only commercial producer of lavender left in Britain. With medicinal and perfumery demands falling and cheaper imports of lavender rising, many of the traditional growing areas in Britain saw their distinctive blue fields shrink away and with them, a rich vein of local culture.

In the 1953 celebrations for the coronation of Queen Elizabeth II, Hitchin in Hertfordshire had a special claim to fame because the town's most famous export – lavender – had been cultivated there since the time of the first Queen Elizabeth. Like Mitcham, Hitchin had been renowned for its lavender fields since the early nineteenth century when enterprising pharmacist Edward Perks inherited a local apothecary business from his father and, like Potter and Moore, embarked on large-scale lavender planting, harvesting and distilling, and on making Perks Lavender Water a household name. By 1952, though still crucial to the town's sense of identity, the heyday of Hitchin lavender was over. Within a decade, the old pharmacy in the High Street was demolished and in its place came a modern Woolworth's store. The new shop apparently did not quite succeed in erasing its predecessor, because customers in search of stationery, school socks and sweets often reported a strange smell of lavender. There were even sightings of Victorian ladies in crinolines ascending an old staircase that no longer existed. Few things are as unpredictable as retail habits and consumer tastes though, for now Woolworth's is a thing of the past and lavender, a rising trend.

Lavender farms are flourishing once again and, far from being the exclusive preserve of farmers and pharmacists, they are welcoming people from far and wide. The fields I stopped to visit on that hot summer heart-of-England morning were certainly full of folk. The surging drifts of blue were scattered with figures: everywhere heads were popping up like mermaids, apparently bent on getting the perfect angle for photographing a family member caught in full summer colour against the deep blue hillside. One young woman, arms outstretched in a saffron dress, stood out like an EU star spinning away from the rest of the circle. English lavender fields are now firmly on the route for tourists – who can stop at specialist farms not only in Hitchin, but also in Gloucestershire, Kent, Norfolk, Somerset, Surrey or Yorkshire. These are the English answer to the well-established Routes de Provence, which take coaches, cars and cycling parties through the warm, lavender-lined lanes of southern France. The French harvest falls from late July to the end of August, with numerous local festivals dotted through the month.

There is something very inviting about these hidden hives of sensory delight. Unlike any other crop, lavender lies in long, straight stripes of blue boot-brush bushes. Once the plants reach full height, there is no obvious sign of where one stops and the next begins. Seamless lines of different varieties, from royal blue 'Munstead' to paler 'Blue Ice', from dark purple 'Hidcote' to pink 'Loddon', white 'Alba' or 'Arctic Snow', create contrasting, strangely nautical patterns, stretching across the fields. Closer inspection reveals neatly symmetrical hemispherical plants, with flower-heads bursting out like controlled explosions on slim, pale green stems. Each perfectly balanced mascara wand has a mass of tiny flowers, which open one by one into a little lilac torch. As warm breezes shimmer through the flowers, the unmistakable scent drifts up like a heat haze. It is when you pick a flower and rub the tiny ruff at the base that the pungent shock of lavender really hits home. However familiar this flower may be, it still has the power to startle, as its hidden intensity bursts out.

If the lavender industry once meant bending over the crop to cut the stalks by hand, risking sunstroke and multiple bee stings, modern farmers can now employ a purpose-built lavender harvester with heavy-duty steel choppers, toothed like a stag beetle, which slices relentlessly through the bristling rows, gobbling up the flowers and expelling them behind into a tall trailer. The cut crop is fed into the distillery, where rainwater, pure and soft, is heated until the steam bursts the glands in the lavender neck, releasing the essential oil. The process is not so different from the practices that have sustained Mediterranean lavender growers for thousands of years, but very new to the lavender industry of contemporary Britain are the tourists and trendiness. While a revival of palsy drops or lavender lice applications seems very unlikely, lavender has found a new fan base in contemporary kitchens. It is currently in vogue for all sorts of culinary purposes, whether being folded into stiff meringue to make lavender pavlova, pounded into sugar and baked into lavender biscuits, rubbed into lamb chops before grilling, or adding a certain *je ne sais quoi* to roast parsnips. At the night market at San Marco in Florence last September, I saw huge sacks of seeds labelled 'Lavanda della Provence' under piles of blue-grey-headed bunches of dried stalks. Fresh lavender flowers macerated in spirit also make one of the most popular American gins, while fresh sprigs and dried flowers are in demand as botanical garnishes for many of the newer blends. Modern lavender appeals to cutting-edge cooks, bibbers and consumers, who still like to feel some connection with an older, more grounded lifestyle.

In China, lavender is increasingly sought by young lovers at Qixi – the Chinese equivalent of Valentine's Day. The festival takes place on the seventh day of the seventh lunar month and commemorates the ancient astronomical myth of the Cowherd and the Weaver Girl, separated for eternity except for one day in the year, when the Milky Way bridge built by magpies allows them to meet for a few hours. Traditionally, this was a day when newlyweds would exchange love tokens, while unmarried women, hoping to attract a spouse, demonstrated any feminine skills they might have. Nowadays people are more likely to send text messages and bouquets, or, perhaps, visit a lavender field. Beijing has a lavender theme park called Blue Dreamland, while Chongming Island in Shanghai boasts a lavender and egret love park. Nearby restaurants advertise special dinners, with dishes such as lavender crispy duck on the menu. The more adventurous or affluent may travel as far as Tasmania, which is now a world-famous lavender centre, thanks largely to a small purple bear.

When the enterprising owners of the long-established lavender farm on the Bridestowe Estate in Nabowla began stuffing spare dried lavender into lilac-coloured, lavender-scented teddies, they had no idea what an international celebrity Bobbie the Bear would become. Once the beautiful Chinese actress Zhang Xinyu had posted a photo of herself cuddling the little purple teddy, demand for look-alike lavender bears soared. Even the Chinese president, Xi Jinping, and the first lady were presented with their very own Bobbie Bear while visiting Tasmania. For a time, the booming Tasmanian trade was threatened by twin bears now being manufactured in China, so the Bridestowe Estate created Lavender Bear Productions and cast Bobbie as the star of a new series for children. Among other gentle adventures, he drives a tractor and shows people how to harvest lavender.

Lavender teddy bears may stir memories of the old nursery rhyme, 'Lavender's blue dilly dilly, lavender's green', which has comforted generations of new parents by helping to soothe wakeful babies into peaceful slumber. Originally this little poem was not quite so innocent. When first published in the seventeenth century, 'Lavender's Blue' (or 'Diddle, Diddle') was a love poem for adults, carrying a fairly frank message. In early versions of the song, the familiar verse continues:

> *I heard one say, diddle diddle,*
> *Since I came hither,*
> *That you and I, diddle diddle,*
> *Must lie together.*

Lavender pillows, it seems, were never intended just to send you off to sleep.

All this may come as something of a surprise to anyone whose idea of lavender has long been preserved in little lacy bags and linked to elderly aunts. Lavender has been the companion of older ladies since at least the eighteenth century, judging by William Shenstone's formidable *Schoolmistress* and her garden full of lavender spikes, though it was probably Charles Dickens who did most to reinforce this idea. His older female characters tend to favour lavender-coloured dresses, and, in an early *Sketch by Boz*, the home of the Crompton sisters is marked by 'a strong smell of lavender', as if its inhabitants are attempting to avoid decay. Queen Victoria's penchant for lavender probably did little to help its image, as she steadily turned into the most famous elderly lady in the world.

The year after Queen Victoria's death in 1901, Myrtle Reed published a novel called *Lavender and Old Lace*, in which the heroine stumbles on an ancient, once white, lace wedding trousseau in her aunt's attic and is almost overwhelmed by the smell of lavender. 'Lavender and old lace' meant old-fashioned, genteel, carefully preserved and often a little sentimental: William Makepeace Thackeray doubted whether any woman 'however old' had not kept her wedding clothes 'stowed away, and packed in lavender, in the inmost cupboards of her heart'. Lavender was often seen as the preserve of spinsters, too, a plant that grew best in 'old maids' gardens'. It was the perfect plant for the American playwright Joseph Kesselring to evoke in his spoof murder mystery *Arsenic and Old Lace*. The two spinster sisters, living in Brooklyn and passing their time poisoning old men with their home-made wine, made a great splash on Broadway in 1939, before becoming world famous through Frank Capra's screen adaptation, which starred Cary Grant as their hapless nephew. Successful films sometimes change popular perceptions, but the purple posters featuring the grand thespian dames, Judi Dench and Maggie Smith, seemed to set the seal on the plant's popular associations with single ladies of a certain age: *Ladies in Lavender* – what else could they be in?

Cole Porter played on the stock figure of the slightly frustrated older woman in a song about a famous gigolo, with 'just a dash of lavender' in his nature, in the 1929 Broadway musical, *Wake Up and Dream*. The impoverished young man's preference for ladies who are 'wealthy rather than passionate' is an old joke, but the popularity of 'I'm a Gigolo' probably helped spread the idea that gay men were born with 'a dash' of lavender. By the McCarthy era of the 1950s, this was no longer a gentle joke. The clamp-down on anyone suspected of communist connections is very well known, but the persecution of gay men, referred to contemptuously as 'lavender lads', has only recently been recognised. Whether lavender was too far across the spectrum from the red-blooded post-war American male, whether the scent of eau de toilette triggered lurking anti-Gallic suspicions, or whether the lavender scare grew from another root entirely is hard to tell – whatever the case, it was decidedly suspect in certain quarters. Far from being the sole preserve of single ladies of a certain age, this sweet-smelling flower, which flourishes on warm stone walls, has also been picked as a figure of resistance to prejudice. In the 1980s a lesbian and gay community bookshop opened in Edinburgh under the sign 'Lavender Menace'. Its quarterly newsletter, 'Lavender Lesbian Lists', was aimed at 'lesbians and other irreverent women' – no 'ladies' here.

However, the ageist and sexist association of 'lavender and old ladies' is perhaps not as firmly rooted as it has sometimes appeared. As singleness in later life ceases to be pitied, or feared, or quietly held in contempt, lavender may come to be universally admired as a sign of independence and resilience. Dorothy Parker, who knew a thing or two about prejudice, sent up the prevailing attitudes to prim spinster aunts in her witty lyric, 'The Little Old Lady in Lavender Silk'. The little old lady, having turned seventy-seven and just about prepared to admit 'I shall shortly be losing my bloom', decides to assess her life to date:

> So I'll say, though reflection unnerves me
> And pronouncements I dodge as I can,
> That I think (if my memory serves me)
> There was nothing more fun than a man!

Dorothy Parker, in common with so many of life's great survivors, knew very well that little old ladies had not always been old and that, in any case, age brought freedoms of its own.

The capacity of 'Sweet Lavender' to keep its colour whatever the weather and preserve its fragrance over many years makes it an emblem of lasting resistance to whatever the world may fling in its direction. This flower may seem to belong in the margins or merging among the masses of more eye-catching blooms, but its staying power is second to none. Through the ages, it has proved a most enduring and adaptable plant, appearing in medieval medicine chests and ornate Elizabethan knot gardens, filling entire lawns of elegant country houses, inspiring the informal cottage garden styles at the turn of the twentieth century and the immaculate, geometric designer gardens of the twenty-first. Its traditional associations with ageing really mean that lavender carries the secret of a long, healthy life.

'But how do you imagine a gillyflower?'

Gillyflowers

The very name conjures up hot summers and heady scents, bright colours and clouds of bloom, crumbling walls and gardens hidden away from the bustle of modernity, almost gone yet ready to reappear in fleeting, unframed memories of a time long lost. Charles Ryder, looking back on Oxford before the Second World War remembered gillyflowers growing under the window of his rooms during the summer term, in the season of punting, picnics and college balls. In those days, students seem to have been remarkably free of anxieties about exams and CVs, at least in the pink-tinted portrait so beautifully evoked in *Brideshead Revisited*. Gillyflowers seem the very essence of a summer's dawning, when the temperatures begin to rise into the steady warmth that makes going out deliciously spontaneous, coatless, hatless and umbrellaless. And in the 1920s, there was none of the alternative armoury of sun-tan cream, insect repellent, bottled water and sunglasses. The fragrance of gillyflowers wooed people out of doors into sunny carefreedom and wafted the summer in through wide open windows. But how do you imagine a gillyflower?

These flowers are very rarely listed in the index of a modern gardening book, though they may appear as the alternative name for officially sanctioned species. The Royal Horticultural Society includes 'gillyflower' as a common term for stocks, but also mentions that the pink is sometimes known as a 'gilly flower', the wild ragged robin as 'cuckoo gilly flower', and the wallflower as 'wall gillyflower' and 'yellow gillyflower'. The *Oxford English Dictionary* widens the field much further, listing 'African', 'castle-', 'clove-', 'Dame's', 'English', 'garden', 'marsh', 'mock', 'Queen's', 'rogue's', 'sea', 'stock', 'striped', 'Turkey', 'water', 'Whitsun' and 'winter' as various types of gillyflower, as well as the 'gillyflower apple' and 'gillyflower-grass'. With such a huge floral array, relating to so many different species, we might begin to wonder whether *gillyflower* really just means 'flower'.

When in doubt about anything floral, I tend to ask my mother, who was unhesitating in reply: 'A gillyflower is a wallflower.' Since my mother's name is Gilly, her view carried even greater authority than usual. The indefatigable Mrs Maud Grieve, whose splendid *Modern Herbal* is a favourite cornucopia of botanical riches, agreed: 'gillyflower' is a synonym for wallflower. As this is a key identification in the *Oxford English Dictionary*, too, set well above the bewildering list of sub-gillyflowers, I would be happy enough to go along with it, henceforth considering any wallflower I might encounter a gillyflower and imagining Charles Ryder gazing from his ground-floor rooms above a haze of deeply scented saffron, orange and rust-red early summer blooms. The idea of a wallflower is, after all, in keeping with the situation of the young first-year student, who, like all the 'Gentlemen who haven't got ladies', is requested to vacate his college on the day of the ball. Except that wallflowers are not the only flowers identified by the *Oxford English Dictionary* as gillyflowers: this word refers to stocks and pinks too, just as the RHS, and indeed quite a number of other experts, have suggested. Geoffrey Grigson, compiling *The Englishman's Flora* in 1958, considered gilly-flowers to be stocks, while Richard Mabey, taking on the vast task of creating *Flora Britannica* more than thirty years later, described the clove pink, *Dianthus caryophyllus*, as the 'Tudor "gillyflower"' and 'chief ancestor of clove-scented pinks and carnations'. Stocks and pinks cannot be set aside,

Flowers formerly known as 'gillyflowers' and also known as carnations and wallflowers.

even on the grounds of family affection, in the quest to find the true gillyflower. And yet, there is something oddly satisfying about being unable to attach the half-magical name of the gillyflower to a single, recognisable species and retaining a sense as gently layered as the petals of the various plants it may evoke.

There are some ingenious explanations for the perennial uncertainty. *Gillyflower* probably derives from the Old French word for 'clove', *girofle* or *gilofre*, which grew into the anglicised *gillofer*, *gilloflower*, *gillyflower*. The English word for the spice came from *clou de girofle*, the nail-like seeds, which gradually became known as cloves, while the *girofle/gilofre* developed separately into gillyflowers, the plants with an aromatic clovelike scent, including clove-pinks, or carnations, stocks and wallflowers. Just as things seem to be making sense, however, another possibility pops up. The garden historian and medievalist John Harvey suggested that *gillyflower* may derive from the Spanish *alhelí*, instead, a word used for both wallflowers and stocks, which was imported from the Arabic *al-khairi*. And so a perfumed cloud descends again over the gillyflower's mysterious past. Since gillyflowers, whether in the shape of wallflowers, stocks or pinks, have been growing in Britain since the Middle Ages, it is not surprising to find more homely roots for their name. A Regency recipe for carnation syrup requiring a pound of 'Clove July-flowers' suggests that gillyflowers are called after the flowering season, as Dr Johnson (who generally preferred the home-grown) concluded. And yet this line of argument was not so easily grafted on to the earlier incarnations of the gillyflower as *gilofre*, *gillofer* or *gillyvor*. Whatever the derivation, there is little doubt that gillyflowers were spicy-scented. In some areas this meant wallflowers, in others, stocks or pinks. This chapter begins with a blank space rather than an image of a flower, so that readers can draw their own gillyflower, like a carnation, stock, or wallflower, or as like some other imaginary plant as they dare.

One person's wallflower is another person's pink is another's stock: our ideas of summer are at once individually and locally coloured. For those in Lincolnshire, for example, gillyflowers are vigorous flowers that brighten beds or burst from stony crevices in orange and gold, raw umber and red. Their warm, bright colours appealed strongly to the early twentieth-century artist George Taylor, whose overflowing bowls of thick oil-painted wallflowers hang in the Usher Art Gallery in Lincoln. But these flowers grow easily in many areas, surviving on thin soil, thriving in full sun. They are common in rockeries, walled gardens and yards and crop up in quarries, ruins or wasteland. They are carefully cultivated in herbaceous borders and grow like an unkempt ginger fringe, often streaked with purple and mauve, along the chalk cliff tops of the south coast or the hot rocks of the Mediterranean. These are the 'castle-gillyflowers', often spotted spilling from the crannies of dilapidated abbey walls or antique parapets. Walter Scott fondly remembered the wallflowers sprouting from the shattered tower at his grandparents' home in the Scottish Borders, though he did not call them gillyflowers.

The wallflower's ability to cling on tight in the most unlikely places has given it a reputation for tenacity and connotations with the kind of love that alters not. Robert Herrick's poem 'How the Wall-flower came first, and why so called' offers a romantic myth of origin, in which a 'brisk and bonny Lasse' falls so passionately in love that she scales her prison walls using a silken string, only to find that it is not, alas, quite up to the task. The string breaks, she pitches to her death, and is then transformed into a golden wallflower. Since in another lyric, Herrick likens first kisses to 'Gelly flowers', he seems to have kept wall-flowers in a separate tray from gillyflowers in his imaginative seed cabinet. William Morris, on the other hand, was clearly thinking of wallflowers for the medieval tournament in his poem 'The Gilliflower of Gold'. The scarlet heart of the yellow flowers, like the petals of the rich red wallflowers, often inspired thoughts of blood. In Morris's poem, a determined knight is driven by thoughts of his lady, 'Bow'd to the gilliflower bed, / The yellow flowers stain'd with red'. As he stirs himself to battle against his opponent, he chants what may now seem an unlikely refrain: '*Hah! Hah! La belle jaune giroflée.*' For the nineteenth-century designer of fabrics, furniture, poems and paintings, the gillyflower was noble, heroic and a quint-essential element of the intensely vibrant world that he longed to recover for the modern nation of industry and mass production.

These flowers have always evoked an earlier era, but what fascinates some is for others merely passé. The cast-off coats that ended up in the old clothes emporia in Monmouth Street were known as 'wallflowers', presumably because they were left hanging on hooks, their once bright colours now somewhat faded. If gillyflowers evoked a magical past, they could also mean second-hand and outdated. For John Clare however, creating a rich pastoral portrait full of 'Old fashiond flowers' in *The Shepherd's Calendar*, gillyflowers were among the glories of June. Here the 'single blood walls of a luscious smell' are side by side with 'white &

purple jiliflowers', which suggests that he was both gathering together and distinguishing between wallflowers and stocks. As stocks belong to the same botanical family as wallflowers – the *brassicae*, or cabbages – they were natural companions in poems and gardens. On the other hand, he may have been talking about carnations.

Carnations, pinks or clove gillyflowers are often variegated, striped like purple, red, or maroon and white rosettes, with petals outspread and pinked. Though these flowers can be the colour of crushed strawberries blended into natural yogurt, the name 'pink' predates that of the colour and probably refers to the thin zig-zag edging of the petals. Wild carnations or clove pinks were generally deep red and valued as a food colouring as well as for their strong fragrance. They were known by the Elizabethans as 'Sops-in-wine' because they floated in drinks like wafer-thin croutons. By the early twentieth century, their distinctive colour and shape conjured up cans of condensed milk and contented cows more readily than wine, thanks to the runaway success of the Carnation Dairy plant near Seattle. The red carnation soon became a symbol of Soviet Russia, too: I remember seeing huge artificial carnations being hoisted up for Victory Day in a square in St Petersburg, when it was still Leningrad, in the late 1980s.

Carnation petals unfold into single, double or multiple layers, some so puffed out that their anthers are completely hidden. This is why they appeal to those keen on crafting artificial floral decorations, as the heads are so easily fashioned from tightly gathered tissue or a red paper napkin. When real flowers were harder to come by during the winter, a hand-crafted red carnation might just about do for a Valentine. Nowadays, fresh carnations are a big business crop, cultivated on a grand scale in California, Kenya, the Netherlands, South America and Spain. Efforts to combat the drug trade in Columbia have included the promotion of peaceful horticulture, and the country is now one of the world's largest producers. In the 1930s, Denver was the international 'Carnation Capital', supplying carnations for the coronation of George VI from huge greenhouses in Colorado. These flowers keep surprisingly well when cut and, light as feathers, can be flown wholesale across the world.

Red, pink or white carnations often adorn the black jackets and flapping gowns of Oxford students heading for their Final exams. They are traditional adornments of morning suits, though currently roses seem more popular as boutonnières for the fashion-conscious bridegroom. Carnations were famously the flower of choice for Oscar Wilde, though he sported a flower as green as the stem. They are flowers for special occasions – acceptable whatever the wearer's gender. And for Wilde, all occasions were special if graced by his stylish presence.

Wilde, who knew his Shakespeare well, was aware that in *The Winter's Tale*, carnations share with gillyflowers the accolade of 'fairest flowers o' th' season'. The season in question was the sheep-shearing season – or early summer. Since Perdita describes gillyflowers as 'streaked' and mentions that they are sometimes called 'nature's bastards', she is understandably wary of letting any into her garden. Shakespeare was playing on the parallels between the human world, with its prejudice against unlawful cross-fertilisation, and the more anarchic realm of flora, where the natural hybridity of gillyflowers or *Dianthus caryophyllus* produced a riot of different colours, stripes and fancies.

Wild clove pinks grew on walls just as easily as wallflowers:

Oscar Wilde sports a green carnation for the opening night of
The Importance of Being Earnest.

both plants originally seem to have crossed the Channel as seeds hidden in stones from Normandy which were imported to build castles for William the Conqueror. Their matching qualities of brightening keeps and crenellations contributes to the perennial difficulties in picking out the true gillyflower from the mixed bunch of possibilities. Unlike the strong, often woody stalks of wallflowers, the stems of pinks are more reminiscent of an elegant and elaborate plumbing system, each smooth, cylindrical section slotting into the next, apparently secured by a long green wing nut, with little pipes decreasing in size at increasingly divergent angles as the plants grow. Their buds are smooth and tight as grenades, until they crack open in the softest explosion of pink and white, peaches and ice cream, plum and cream, purple, crimson and mauve.

Stocks rise more surely from flower-beds than carnations and pinks, especially if their side shoots have been trimmed, allowing the central stem to grow taller and stronger. A healthy bed of stocks stands around in multicoloured finery like a convention of feather dusters, and can last from June to September. Stock gillyflowers, *Matthiola incana*, also go by the rather less attractive name of 'hoary stocks', because of their strong, rough-textured stems. They may be native to the south coast of Britain, though they originate primarily in southern Greece and the Mediterranean. It is the night-scented or evening stocks, *Matthiola biflora*, that carry the most pungent whiff of a settled summer. And these gilly-flowers have spread around the globe. At the heart of the Butchart Gardens on Vancouver Island is the Sunken Garden, a great mass of pink, lilac, purple and white stocks which suffuses the sea air with scent; in New Zealand, a flourishing bed of purple stocks surrounds a choir of cherubic statues in the Botanic Garden in Christchurch.

During the eighteenth century, when stocks really came into their own, the most prized were the red and purple flowers (*Cheiranthus coccineus*), bred by the highly successful horticultural partnership of George London and Henry Wise at their nursery in Brompton Park, South Kensington (now the site of the National History, Science and V&A Museums). Brompton stocks became famous across Europe and seem to have done their bit for Anglo-French relations in an age strained by recurrent warfare. The influential landscape gardener Henry Phillips, who did so much to transform Brighton and Hove in the early nineteenth century, endured a rather trying trip through Normandy with a travelling companion who had never before set foot outside England. Nothing could please the reluctant tourist, who found the French soup 'meagre', the peas 'sweet', the wine 'sour', the coffee 'bitter', the girls 'brown', the petticoats 'too short', the roads 'too straight', the inns 'dirty' and the language 'unintelligible'. His unhappy *vacance en France* was finally saved, however, by the chance discovery of a garden full of stocks, which reminded him of his Sussex home. When told that these were *Giroflier de Brompton*, he immediately ordered champagne to share with the owner of the garden and her family, before going on his way with a sparkle in his eye, a sprig of gillyflower in his buttonhole and a new motto for life: 'Thanks to the Brompton Stock.'

Whatever their local character or name, any garden is enriched by gillyflowers, sprawling out, clinging on, or rising straight. Gillyflowers belong to an older world, or rather worlds, where daily life operated according to local norms as much as national standards. The changing weather registers especially strongly in rural communities, and so to name plants by their flowering season is perfectly logical. This way of thinking leaves traces in the popular, often regional, names and makes sense of calling a stock or carnation a 'July flower', irrespective of etymologies. Since Culpeper stated that wallflowers bloom in July as well, the gillyflower reminds us that the habits of plants are perpetually changing and what seems normal today might surprise our ancestors. Old flower names often carry memories of local stories and beliefs, of unofficial ways of understanding the world. To call a wallflower or a carnation a gillyflower is not to reveal ignorance, but familiarity.

The gillyflower's identity traditionally depended on its

scent, which offered an alternative way of grouping plants from the standard system of modern botanical classification that is based on observation of plant structures – the number of sepals and petals, stamens and styles, the shape of the calyx and the corolla, the leaves and the length of the stalk. Colour and texture can be important botanical markers, but fragrance, so hard to describe, often only gets a mention if it is especially strong. The idea that 'gillyflower' covers a variety of spicy-smelling blooms is an older, pre-Linnean way of thinking about flowers, a way that made sense to those like Francis Bacon, who knew 'what be the flowers and plants that do best perfume the air'. Gillyflowers belong to a world of multisensory experience, of smells as well as sights and sounds. They affirm the power of fragrance to translate us into a state of heightened experience and to find a permanent place in the memory. The earliest childhood memories are often olfactory, and so distinctive smells have a special power to trigger vivid thoughts of vanished things. What gillyflowers really mean is not a particular species of plant, but the half-forgotten world of childhood and times almost out of reach. Gillyflowers grow in the gardens that everyone wants to recall and reinhabit, and yet cannot quite bring into focus. They lie in the mind, brightly lit and yet oddly indistinct. No wonder we still long to find the true gillyflower, the plant that promises the key to the walled gardens of lost time.

*When John Evelyn's famous book on trees, Sylva, was enlarged
in 1776, it was packed with beautiful new engravings including
'The Lime Tree'.*

Lime Flowers

During the recent restoration of Trinity College chapel in Oxford, three centuries of wear and tear, dirt and layers of paint and varnish were painstakingly stripped away from the Grinling Gibbons carvings. As the years were gently removed, the shapes of the wood sculptures began to lighten and clarify into the perfect petals, stems and leaves of his original vision. These white gold arrangements of flowers, so delicate you can almost see the light streaming through the petals, were created from lime wood, the strong, smooth, finely grained and wonderfully malleable timber that has long been the woodcarver's favourite. The flowers include exquisite tulips, lilies, daisies, vines and leaves, all as carefully observed as a Dutch still life. And every one of these wooden blooms is, in a sense, a lime flower. Among the highly controlled floral cornucopia are small signature clusters of stars and forget-me-not shapes, which might be a tribute to the trees that furnished the great wood sculptor with his golden trove of raw materials. When lime – or linden – trees flower in high summer, there is a pale explosion of tiny white filaments, which spread out like multiple shatterings in a pane of glass; just prior to this exuberant profusion, as the spherical buds are opening, the sepals spread in rounded rings, which then slim into stars. As he made the flowers permanent through his art, Gibbons also caught the botanical cycle of life, budding and blooming and turning to seeds, which might eventually grow into saplings, mature trees and one day into some beautiful wooden creation.

Lime trees are a common enough sight along roads and paths, in parks and woods across Britain, Europe and the United States (and known variously as lime, linden, basswood, *tilleul*, *tiglio*, and botanically as *Tilia*). These tall, soft, swelling cones of gently spreading green are not always as obvious in a line of broadleaves thick with early summer growth as some more showy species. Their elegant trunks and branches are rapidly covered by cascades of leaves in the shape of asymmetric hearts, like the doodles in a school book, with the lovers' initials left blank and perhaps encrypted in the faint, indecipherable runic lines which cover each leaf. But it is only when you see limes at close quarters that these details are visible: in a more distant parade of trees, they are not especially noticeable, except for the pale green highlights that suddenly appear in their choppy canopies during June. At the tips of the branches, where they ramify into twigs, pale, slim, leafy bracts appear, arching away to admit the light and allow fresh air to circulate around clusters of tiny spheres, which stick out at sharp angles from the stalk. It's as if a shower of quotation marks and asterisks and parentheses has fallen from somewhere overnight. These flower-buds hint of silences and ellipses, of things almost being said: the summer secrets of the lime trees stop with rows of dangling dot, dot, dots. At the height of midsummer, when the night is in deepest retreat, these cloaks of repeating crescents and butter-coloured flower clusters suddenly burst into full glory as the sun's warmth turns to heat.

Any thought that the flowers might be too small to make much impact on an abundant wood or garden of bright kaleidoscopic blooms is instantly overcome by their rich, honey fragrance. The Victorian poet, Matthew Arnold, lying in the long grass at high summer, where 'air-swept lindens yield / Their scent, and rustle down their perfumed showers / Of bloom', was wafted away on the thick breeze to the lost world of the Scholar Gypsy. Few sensations trigger memories as powerfully as a distinctive scent and few trees exude scents as evocative as the lime. In the 'Overture' to Marcel Proust's *À la recherche du temps perdu* (*Remembrance of Times Past*), the taste of a small scallop-shaped madeleine cake dipped in tea makes the narrator's mind flood with memories of the cake soaked in lime-flower tea that his aunt had given him long ago, and with it all the flowers in the gardens and parks and rivers of his childhood. Dried lime flowers infused into a cup of boiling water create a soothing, aromatic drink. Lime flower tea, or *tilleul*, as it is known in France, is an old remedy for anxiety and nervous symptoms such as palpitations, insomnia and indigestion, so it is often associated with comfort and loving care. Those subject to fits of hysteria were encouraged to take long deep baths in hot water infused with lime flowers. In Germany, the flowers of the lime tree also have long-standing links with love, but especially illicit love. In the famous medieval song 'Unter der Linden', the only traces of the lovers' tryst are the crushed flowers on the ground beneath the sweet leaves of the lime tree. The sensuous secrecy of such associations makes Samuel Taylor Coleridge's forlorn evocation of his own loneliness in 'This lime-tree bower my prison' all the more powerful.

In late July, Green Park in central London has a heavy, honeyed atmosphere, as the lime flowers waft across the hot afternoon. Their scent is easier to recognise than articulate. Joseph Conrad understood the lingering yet elusive power of lime flowers when he described the voice of his enigmatic character Señor O'Brien as possessing 'the faint, infinitely sweet twang of certain Irishry; a thing as delicate and intangible as the scent of lime flowers'. In fact, these flowers often seem to have a voice of their own: a steady hum, growing louder as more and more of the tiny buds burst. The first of a swelling cluster to open is often utterly obscured by the much larger, darker and more densely filamented form of a bee, balancing on the tiny stem in a desperate attempt to get at the nectar. There is nothing new about the insect's predilection for the heavenly scented flower clusters: Roman authors were well aware of the annual bee-lines for the lime tree. European and American beekeepers still routinely plant the trees around their hives to maintain a supply of valuable lime flower, or linden, or basswood honey, or *miele di tiglio*. Traditionally, lime flowers were regarded as the source of the finest honey, a pure gold whose flavour intensifies as it lightens and crystallises. This is the sweetness and light carried home by bees from their busy summer visitings.

For anyone who associates limes most readily with green fruit, a sharp citric taste, or a very hot pickle, this abundance of sweetness may come as something of a surprise. Lindens keep many secrets, but one of the biggest is that they are not related to the family of trees which produce the citric fruit called limes. The tradition of referring to the *Tilia* as the lime is probably the result of an acoustic error, as the Germanic *linden* became 'lind' or 'line', which sounded like 'lime' to British ears. Although the pale green summer bracts of the *Tilia* may remind some observers of the citric slices floating in a glass of gin, the small, round, underripe lemon-like fruits come from an entirely different botanical species. The source of this fruit (and the related nickname of scurvy-free British sailors or 'limeys') is the *Citrus latifolia* or Persian lime. Citrus trees are a large family with different branches around the globe, from Australian

The late seventeenth-century wood sculptor Grinling Gibbons turns lime wood into flowers for Trinity College chapel, Oxford.

limes to Key limes, Kaffir limes, Persian (or Tahiti) limes and Rangpur limes. None of these are remotely related to British lime trees, though many have pretty, white star flowers of their own, a little larger and less secretive than the creamy clusters of their European namesakes. When Coleridge's nephew, Henry, visited the West Indies in 1825, he was struck by the difference between the flowers, fruit and foliage of Montserrat and those of his Devon home. The natural greens were quite unlike those of England and soon the island's orchards became established as the main source of Britain's citric limes and lime juice. When people today refer to 'lime green', the name of the colour derives from the fruit that grows on *citrus* trees, but there is nothing to stop us thinking of the paler buds and bracts of the *Tilia*, too.

John Ruskin's portrait of a much admired milk thistle, 'Fat Fitie'.

Thistles

It's hard to go very far in Scotland without coming across a thistle. Anyone arriving in Edinburgh by train will probably spot quite a few before reaching the top of the Waverley Steps. Heading in the opposite direction towards Holyrood or Edinburgh Castle, the Royal Mile is all a-bristle with purple-headed tea towels, T-shirts, mugs and postcards. You can buy shortbread thistle biscuits and thistle tea, recommended for the flavour rather than the texture. Across the top of the *Scotsman* newspaper, the thistle stands proud. There can be no doubt about which flower is Scotland's national bloom. The round head with the tufty crown is instantly recognisable, perhaps only mistakable for a pine-apple, which would be an unlikely choice in this northern capital. But what is the enduring appeal of this spiky wild flower?

In garden guides thistles are generally treated as among the toughest of weeds, prompting alarmist paragraphs on how easily they seed and how very difficult their fiendishly sharp leaves and impossibly long taproots are to remove.

The wild flower's negative image goes back to the beginnings of human history – at least according to the Old Testament. After Adam and Eve have eaten the forbidden fruit, the ground is cursed with bearing 'thorns and thistles'. As if this wasn't enough, Job in the midst of his catalogue of troubles foresees thistles growing in place of wheat, and then in the New Testament, when Christ warns his followers how to spot someone's true character, this is the flower that pops up: 'Ye shall know them by their fruits. Do men gather grapes of thorns, or figs of thistles?' Biblical bad press usually proves a difficult legacy for any natural phenomenon, so the thistle's triumph over such odds is greatly to its credit.

Robert Burns did his bit for the thistle. In a poem recovered from his papers and published soon after his death, he wrote 'To the Guidwife of Wauchope House', in celebration of 'The rough bur-thistle spreading wide / Amang the bearded bear'. Although most farmers would have no qualms about removing such an invasive weed from their crop of barley, Burns declared proudly that he threw aside his 'weeding heuk' to spare Scotland's 'symbol dear'. The 'burr thistle' is the Ayrshire name for the spear thistle (*Cirsium vulgare*), also known as the 'boar thistle' across the south of England, because of the thistle-like bristles running down the back of a wild boar.

The thistle's appeal to patriotic Scots was boosted by Sir Walter Scott, who was not only very taken with Burns's poem, but also a master of image-making. Scott ensured that when George IV made his state visit to Edinburgh in 1822 (the first Hanoverian to make such a peaceful overture), the grand procession to Holyrood included sprigs of prickly headed cotton thistle (*Onopordum acanthium*). This national symbolism, however, was by no means a Romantic invention. Thistles had appeared on Scottish coins since the reign of James III in the fifteenth century. The panegyric composed by the poet William Dunbar for the occasion of James IV's marriage to Margaret Tudor in 1503 was an allegory entitled *The Thrissil and the Rois*, which celebrated warming relations between Scotland and England. When James VI united England to his realm a century later, becoming James I and VI after the death of his heirless relation Elizabeth I, his newly minted coins bore the symbol of the rose on one side and the thistle on the other. The symbolism later became a little less clear-cut, after Jacobite sympathisers adopted the rose as their own. England, meanwhile, began to include the thistle in the context of the larger United Kingdom, encouraging thistles to sprout on coins and royal regalia. The Jacobites were keen on thistles, too: the Most Ancient and Most Noble Order of the Thistle – Scotland's answer to the Order of the Garter – was founded by James II and VII in 1687, though the knights' new chapel at Holyrood was destroyed the following year when the king abdicated on religious grounds. The Order of the Thistle was revived in 1703, on the accession of Queen Anne, but another two centuries passed before the grand Thistle Chapel, studded with heraldic devices, was opened in St Giles' Cathedral in Edinburgh.

Far from being a lowly weed, the thistle stands for the oldest aristocratic families in Scotland, whose shields and coronets are ringed with thistle wreaths. And yet the flower's appeal is insistently democratic – as Burns knew very well. This is a flower that needs no careful cultivation, resists the worst of the weather and grows stout and strong on the poorest soil. It stands its ground tenaciously, offering fierce opposition to those bent on its dispossession. The original legend of the thistle plays on the plant's prickly character: a Viking, bent on marauding through Scotland, got such a shock when he stepped on a thistle that his shout of surprised pain gave away the planned invasion and roused the residents to rise and protect their own. Like most good stories, this has little basis in historical fact, but has taken root and grown because of its appeal to national self-image. The thistle's natural defences furnished a long-lived martial metaphor – whether its spikes were seen as spears or swords, maces or arrowheads, bayonets

A Scottish thistle prompts a barefoot Viking to alert the unsuspecting natives to the hitherto secret invasion.

or bombs, or, in more peaceful times, as golf clubs or football studs.

For Christopher Murray Grieve, better known as the poet Hugh MacDiarmid, the national flower offered a multifaceted subject through which to reflect on Scotland's identity. His great poem 'A Drunk Man looks at the Thistle' plays knowingly on Scottish stereotyping while revealing the complexities of Scotland from the inside. There are almost as many facets as there are spikes on a thistle stalk. In Calum Colvin's new portrait of Hugh MacDiarmid, constructed under the gaze of visitors to the Royal Scottish Academy in the last weeks of 2017, the bric-a-brac of modern life combines to create a powerful tribute. When viewed from a certain angle, MacDiarmid's eyes seemed to focus on a pair of tobacco pipes in a thistle jug – the poet is looking at the thistle, the audience is looking at him looking at the thistle while remaining aware of the artist looking at the poet looking at the thistle. The witty construction recognises the miscellaneousness of contemporary Scottish culture, its kitsch and consumerism, debts to the past, seeming coincidences and careful design, while the thistle stands alone, yet surrounded by meaningful clutter. A modern nation is made up of many individuals and objects, somehow collectively forming a recognisable identity. Thistles grow in many places, so Scots may come across this emblem of home anywhere in the world. Whether every – or indeed any – Scot wants a garden full of thistles is another matter entirely.

When it's not being forcibly eradicated by gauntlets, garden forks and weedkiller, the thistle is generally treated with caution. J. H. Crawford, who celebrated the national plant in his late Victorian book *Wild Flowers of Scotland*, described local children peeling back the thorny rind to reveal the 'purple-robed flowers' and consuming them 'with the relish of epicures'. Many food foragers are less convinced. Johnny Jumbalaya's cookbook includes many dishes based on nettles, dandelions and chickweed, but although the thistle does merit a few pages, it is clearly not a favourite. He recommends roasting the flower-buds of spear thistles, but thistle leaves are apparently very bitter if not thoroughly soaked and boiled. There are some brave recipes for 'thistle biryani', 'thistle stroganoff' and 'lamb and thistle stew', but these are for 'those not in a hurry', because the task of removing all the prickles is so laborious.

Admiration for flowers does not normally depend on whether they are nutritious or delicious, of course. It's the look of the thing and its natural habits that ultimately determine whether a flower is welcome or not. The low-level thistles that hide in an untreated lawn do not have much to recommend themselves, but their taller relatives, such as spear thistles or milk thistles, can be a magnificent sight. The grandeur of these statuesque figures, crowned with imperial purple, is well caught in John Ruskin's fine botanical study of 'Fat Fitie'.

Thistles are not inclined to mass along the Scottish border any more densely or magnificently than elsewhere. Only a few hundred yards away from our home in Buckinghamshire is a thick, tangling border of spikes and plumes. There is more than one species of thistle here: a slimmer, rather gangling green plant with needly leaves and a lilac head grows on either side of thicker-set spear thistles, their leaves outspread and stalks like lamp posts covered in barbed wire. Their fat buds bulge like a shoal of puffer fish until at last they seem to turn inside out, ejecting deep purple entrails.

If thistle flowers look just as prickly as the rest of the plant, a careful touch will reveal that the pointed filaments are as velvety as royal cushions. I was disappointed when a great line of spiky, frost-defying, peppermint-cream-coloured thistles was devastated by a herd of young cattle let out to graze in the spring. But even though the thistles had been reduced to forlorn brown stalks, a few were popping up again by June. These are very robust plants and yet, soon after their flowers have emerged in imperial glory, they dissolve into fluff and float away. The soft seeds of thistle are more suited to a fairy tale than a national epic. Unless snaffled up by goldfinches, thistledown flies on the breeze and offers no warning at all of the great spikes that will burgeon from such innocuous-looking seeds.

*Charles Tunnicliffe celebrates the giant sunflowers towering over a
garden in Kent in his striking wood engraving.*

Sunflowers

Who can resist this lion of a flower, with its broad, round face and bright, shaggy mane? The sunflower's buds turn towards the sun, following the movement of the brightest days, while its great green leaves are like an ace of spades played again and again up and down that impossibly long stalk. Each great head is not a single flower but a mass of tiny brown florets with a halo of yellow ray flowers, making it more of a living bouquet. The little button blooms spiral out this way and that, in perfect Fibonacci sequences, like a pair of snail-shell patterns, mirrored and overlaid. Gradually they stiffen into seeds, offering a feast for bullfinches, goldfinches, greenfinches, great tits, coal tits, blue tits and any other birds eager to top up their reserves for the long, spare winter ahead. Golden crowned, bristling with largesse, gazing down on the lawns, the borders and the shorter-stalked blooms, surely this is the king of the garden?

Apparently not everyone is impressed by sunflowers. The Royal Horticultural Society refers to common sunflowers

(*Helianthus annuus*) variously as 'tall, coarse plants' with 'coarse simple leaves' and recommends them as 'Plants for Kids'. Dr Hessayon, whose practical guides are a standby for trowel wielders, regards the sunflower as a 'coarse giant of a plant', best excluded from any self-respecting garden in favour of more compact varieties. Admittedly, as the summer wears out, sunflowers can begin to look a little top heavy, as their golden manes begin to thin and their heads, now disproportionately broad and weighed down with seeds, start to tilt towards the earth. But objections to the sunflower seem to relate more to its natural vigour, judging by Hessayon's weary conclusion: 'these mammoths will be grown as long as children and competitions exist'. These enormous flowers are certainly early favourites, because their very satisfactory black-and-white striped seeds are so easy for small children to pick up and plant. In no time at all a green shoot with propeller-like leaves pops up to show that planting really does work. The shoots soon outgrow the young gardener, rising rapidly like a magic beanstalk, though with a much friendlier giant in view. Planting a dozen sunflower seeds can prompt a guessing game about which will have the longest stem or the widest crown and also allows wisely for a few disappointments along the way.

The pleasure of growing giant sunflowers is by no means confined to childhood. In September 2015, eighty-three-year-old Fred Gibson, who has been nurturing sunflowers in his garden in Hett, County Durham, for six decades, appeared in the national press, dwarfed by his champion flowers. Despite a dismal summer, the flowers had suddenly multiplied exponentially and shot to eleven feet in height. They still did not match the world's tallest sunflower, which had grown to over thirty feet the previous summer in the German town of Kaarst and had to be measured by the local fire brigade. This was an exceptional sunflower, of course, but an ordinary specimen might attain six feet in height and a twelve-inch petal-to-petal span. A row of these along a garden fence will always brighten up August, when so many other blooms are in abeyance. The sunflower's scale and vigour secures its general popularity, but it seems that in the eyes of real gardeners, any plant that grows so easily – and so *tall* – is not one to choose. 'Coarse' sunflowers somehow threaten the delicacy of certain gardens (or rather gardeners), even though they have been cottage favourites for many years.

For such open-faced blooms, sunflowers provoke surprisingly contradictory reactions. Humans have always found – or imposed – meanings on flowers, but sunflowers posed a particular challenge. As natives of the Americas, they were completely unknown in ancient Greece or the Middle East, where so many of the guiding myths of Europe originate. Once these sun-lovers arrived from across the Atlantic, they were still interpreted according to ancient classical or biblical traditions. The botanical name, *Helianthus*, reflects European perceptions of the sunflower's most striking characteristic, as *anthus*, 'the flower', follows *helios*, 'the sun'. The sunflower's heliotropism and the obvious resemblance between the golden ray flowers and sunbeams made it seem the perfect botanical casting for an old tale of devotion to Apollo – no matter if in the original version of the myth, the flower is more like a violet. Ovid tells the story of Leucothoe, a young nymph unlucky enough to fire the obsessive pursuit of the sun god Apollo, who then neglects his devoted admirer, the water nymph Clytie. The sun god,

disguising himself as Leucothoe's mother, gains entrance
to her private room and then to Leucothoe, unwittingly
incensing Clytie. When Clytie reveals what has happened
to Leucothoe's father, he is enraged and condemns
Leucothoe to be buried alive. Apollo is inconsolable and
Clytie is punished by spending the rest of her days yearning
for him, until she turns into a flower, following the sun.
In his striking bronze statue, now belonging to the Tate,
George Frederic Watts caught the moment when Clytie's
body had largely disappeared, leaving only her head, breasts
and shoulders, twisting in anguish as she becomes a
sunflower, yearning for Apollo to the bitter end.

The plant's reputation for heliotropism, or habit of
rotating slowly to face the sun from east to west, also turned
it into an emblem of fidelity and Christian devotion. In
seventeenth-century wedding portraits, sunflowers occasionally appear as symbols of love and obedience. *A Young
Woman holding a Sunflower* by the fashionable Amsterdam
portrait artist, Bartholomeus van der Helst, shows a young
woman, dressed in white, pointing to her heart with her
right hand while holding aloft a sunflower in her left. The
flower, with its promise of abundant seeds, suggests the
promise of loyal obedience and a fruitful marriage, and
perhaps the added attraction of a little worldly wealth. In
a late nineteenth-century book for children entitled
Sunflowers and beautifully bound in a gilt botanical Arts
and Crafts pattern, the writer G. C. Gedge explored a
dilemma facing Victorian girls. The story begins, 'Dear
Cousin Mary, I am a sunflower', in a letter from a young
woman bemoaning her removal from the bright excitements
of urban life to rural seclusion. The story was published
by the Religious Tract Society, so it does not take Mary

long to decide that although some girls, like her cousin, may prefer to devote themselves to the 'sun' of worldly pleasure, she identifies with sunflowers for very different reasons. For Mary, sunflowers suggest 'Christians living eye to eye with God'. The message is clear enough: to identify with a sunflower should mean to follow the paths of righteousness, though there were probably readers who quietly shared Mary's cousin's fondness for city lights.

To those with Christian belief, the sun-following flower meant obedience and religious commitment; to those with a melancholy or even tragic outlook, it meant hopeless or misplaced devotion; to those whose values were more materialistic, sunflowers were associated with wealth and status. Often these flowers inspired strange blendings of tradition and incorporated a few other elements as well. At once godlike and natural followers of God, the stately plants with their large, round heads were open to quite contrary interpretations.

The first sunflower to be grown in Europe is reputed to have held pride of place in the Royal Botanic Garden in Madrid, a wonder of the New World and a massive status symbol for King Philip. Spanish conquistadors came across sunflowers flourishing in Central America and marvelled at a land where even the flowers were golden. So marvellous were these flowers in fact, that the Europeans stuffed the seeds into their pockets and carried them off to Spain, where their sunflowers quickly grew to rival all the other glittering trophies. The plant's astonishing height, hue and habits gave it an air of grandeur, amazing those who had never before seen such a thing. The crowned head that commanded the full glow of the sun seemed charged with divinity, like the king.

The brilliant Flemish artist, Anthony Van Dyck, is famous for his glorious portraits of King Charles I. Given the apparently incessant demands of his patron for more and more, larger and larger, grander and grander royal portraits, it is surprising that he had time for anything else, but the Royal Academy's recent reassembly of Charles I's remarkable art collection included two self-portraits, each featuring a sunflower. In the earlier portrait painted in 1632, the artist poses beside a magnificent sunflower with a fringe of gold as thick as his own hair. With one hand, Van Dyck is pointing at himself, and with the other, at the flower, as if to indicate his personal identification with that symbol of Christian piety and devotion to the king, entirely befitting

Sir Anthony Van Dyck in the earlier of his two sunflower self-portraits, 1632.

the newly appointed court painter. Something about the artist's slightly furtive expression nevertheless suggests that the image may be more complicated than it seems. Perhaps the sunflower, still a relatively rare plant in Britain, had been purloined from the royal garden? Or was it that the artist was having certain misgivings about his new role? The huge sunflower stands right next to his face, like a mirror, necessary of course for a self-portrait but also a recognised symbol of vanity. Van Dyck may well have had qualms about feeding Charles I's self-obsession, given the vast scale of *The Greate Peece*, the royal commission undertaken in the same year, but was there also some unease about his own growing sense of self-importance? The yellow flower accentuates the golden chain around the artist's neck, a gift from his royal patron by whom Van Dyck had just been knighted. This overt sign of favour and material fortune was perhaps a covert sign of bondage. In the later portrait of 1640, the haunted face staring out is in an almost identical pose, lined with anxiety, while the sunflower has moved from the painting to the gilded frame, beaming out above the artist and also forming an integral part of the circular frame: the artist is no longer holding the sunflower, but has been engulfed within it. A year after completing this portrait, as Charles I waged war with his Scottish opponents, Van Dyck died at the age of forty, too early to witness the ultimate fate of the royal master he had painted so many times.

On the vast roof of Versailles, glistening baskets of sunflowers alternate with royal crowns to attract Apollo's beams to King Louis XIV's magnificent palace. These golden forms, silhouetted against the skyline and barely visible from the ground, make attempts to grow the world's tallest sunflower seem rather forlorn. *Le tournesol* was an obvious motif for the Sun King and even the frames of some of the finest paintings in his vast collection, such as Raphael's striking portrait of Castiglioni, were adorned with carefully carved sunflowers. Long after his death in 1715, French craftsmen still drew on the sunflower as an emblem of royalty and the passage of time, creating extraordinary pieces such as the Vincennes clock. Easily mistaken for a vase of flowers, the bronze clock face sits within a golden sunflower at the centre of the porcelain flowers. The Prince Regent was delighted to welcome the Sunflower Clock to Britain in 1819, after Napoleon's defeat had at last made his own position – and the royal collection – much more secure. Three decades before, in 1789, at the start of the revolution that would bring the French monarchy to the guillotine, the great Dutch flower artist Paul Theodor van Brussel had included a sunflower in one of his many paintings of 'Flowers in a Vase'. Here the sunflower was shown falling headlong from the arrangement, its stalk broken.

The sunflower's diurnal rotation led naturally to thoughts about the perpetual movement of time and the turn of Fortune's wheel. One of William Blake's *Songs of Experience* begins 'Ah! Sunflower, weary of time, / Who countest the Steps of the sun'. This enigmatic lyric draws on the mixed traditions of Christian devotion, classical yearning and the ticking clock. Blake's poor sunflower is a devoted follower, now worn out by endless yearning. The youth who 'pined away with desire' and 'the pale Virgin shrouded in snow' have each gone to their graves. Although they may 'arise' and 'aspire', it is not clear whether this suggests Christian souls on their way to salvation or those who have mistakenly denied the flesh and perished unsatisfied. The classical myth of Clytie, which had been adapted to incorporate the sunflower after its arrival in Europe, may have had an influence on Blake's ambivalent flower. Sunflowers were sometimes planted in graveyards, as religious emblems, promising light above the darkness of death or acting as *carpe diem* reminders of the brevity of life. The common sunflower is an annual, so it withers away in the autumn, to rise again only if the seed falls on fertile ground.

The sight of sunflowers growing from graves took on the darkest meaning for the Holocaust survivor, Simon Wiesenthal. After the war, sunflowers reminded him of the harrowing months in a concentration camp, where he had seen them sticking up in a German military cemetery, 'straight as a soldier on parade'. Wiesenthal resented the respect paid to fallen Nazis, whose graves were lit up by the sunflowers, in comparison to the atrocities meted out to his own persecuted people. The fair-haired sunflowers had become inseparable in his mind from memories of a dying German soldier, who had begged for absolution from his anti-Semitic brutality. Wiesenthal's terrible memories were intensified by the horror of the soldier's deathbed confession, the bright sunflowers and his own inability to forgive.

Sunflowers have also offered a brighter perspective to those prone to dark thoughts. When Vincent Van Gogh left Paris for Provence in February 1888, he found a world radiant in deep yellow. Vast fields of sunflowers stretched across central and southern France, like a rough, rippling golden blanket, just as they do today. Van Gogh, exhilarated by the vibrant colours, painted white-blossoming orchards, terracotta cottage roofs, turquoise rivers and scarlet poppies, and could not get enough of the golden sunflowers. He dreamed of an artist's colony in the south of France, where friends such as Gauguin would join him to push the boundaries of canvas and colour under bright Mediterranean light. As a welcome to Gauguin, Van Gogh planned to create 'a symphony in blue and yellow', decorating the studio with 'Nothing but large sunflowers'. Canvas after canvas was coated in urgent orange, chrome yellow, burnt umber, because Van Gogh knew just how quickly the flowers wilted. Exhausted by his painting frenzy and exasperated by Gauguin's portrait of him as a painter of sunflowers, which he thought depicted him as a madman, Van Gogh threatened his guest with a razor before mutilating his own ear. Van Gogh had completed four of his sunflower paintings by September 1888 and worked on further versions before his death two years later. Although he reflected mournfully in a letter to his brother Theo, that 'Pictures fade like flowers', his *Sunflowers* are now among the most admired and highly valued in the world. In 1987, a single painting from the sequence broke all existing art auction records when it sold for nearly 40 million dollars. You can now buy golden T-shirts and pens and coffee sets and spectacle cases decorated with prints of these extraordinarily valuable flowers.

For those who make a pilgrimage to Provence to visit the landscape that inspired Van Gogh, the colour of sunflowers may be intensified by thoughts of his untimely death, but the great, golden fields that still grow there are not planted in honour of their most famous painter. They are grown in France, as in Russia or China, Argentina or India, because of the golden oil and rich nutrients in their seeds. All over the world, if the climate is right and the irrigation well organised, these dazzling, ragged-headed suns flourish in their thousands. The high-energy black-and-white pointed seeds can be eaten raw, or salted, or roasted. They can be baked into cookies, stirred into breakfast cereals, sprinkled over salads, or ground into meal for making bread and cakes. In Germany, a special type of rye bread baked from sunflower seeds and rye makes quite a mouthful – *sonnenblumenkernbrot*. The oil derived from the crushed black seeds, packed with vitamins and minerals, is itself a culinary essential. Though regarded in some kitchens as the poor relation of olive oil, it is less expensive, more widely available and can make a key contribution to chocolate fudge cake. Sunflower oil makes easy work for design and marketing teams, too, because the flowers make such a cheerful eye-catching splash of yellow on labels or packs of oven chips.

Serious production of sunflower oil took off in the 1930s, partly as a result of the Spanish Civil War and its impact on the trade in olive oil. In the seventies, health concerns about animal fats and cholesterol caused sunflower margarine production to rocket. The world's largest producer of sunflower oil is Ukraine, where sunflowers have been a major crop for many years. Among Peter the Great's many legacies was the introduction of mass sunflower planting to eastern Europe. The United States, where these flowers are indigenous, comes in only eighth place in the international sunflower seed stakes. As a single flower-head can produce as many as 2,000 seeds, this is a crop that offers a high return on minimal investment. In Burundi, one of the world's poorest countries, new sunflower farming projects are helping to build a more secure future. The seeds provide highly nutritious food and the yellow ray flowers can be used to dye fabrics. Looking to the future, there are real possibilities for sunflower growth, as its oil is developed as a biofuel. If it ever becomes a serious renewable alternative to fossil fuels, the world's oil wells may become redundant relics one day, as the world fills with fields of giant yellow flowers.

Poppies were valued for their medicinal properties in Renaissance Europe: illustration of the common poppy from an Italian herbal, Liber de Simplicibus by Benedetto Rinio.

Poppies

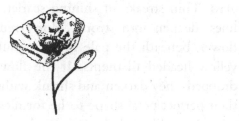

Claude Monet's famous painting of *Les Coquelicots*, or *The Red Poppies*, shows figures in brilliant sunlight moving gracefully through a field waist deep in scarlet poppies. The small boy at the foot of the slope, barely visible above the long grass, holds on to his mother with his left hand, while in his right he clutches a bunch of poppies as red as the band in his straw hat. Scarlet sets off the green of the long grass as surely as the orange pantile roof complements the clear blue sky. This is a vision of natural harmony expressed in complementary colour and light and, more obviously, a celebration of the extraordinariness of an ordinary day. The field poppy, *Papaver rhoeas*, or common poppy, as it is also known, grows wild in the open fields of France, where Monet lived and worked. It is a familiar sight in Britain, too, and across Europe as far as the Middle East and North Africa. This bright summer flower has special appeal for children, when they discover the secret of those intriguingly hairy, hanging pouches, which weigh down the slim stalks as they swell. Small fingers can peel a pouch apart to reveal fold after fold of red petticoat unpacking from its green case. When left untouched these buds soon crack independently into a perfectly circular cup, before opening fully into a broad cross of matching paired petals, pale red and translucent as glass. Thin streaks of shining scarlet, veined with shadow lines, deepen into strong, glossy red at the heart of the flower, beneath the pale green capsule and a gathering of yellow-headed filaments. If the flowers are gathered and dropped, they darken and shrink within a few hours, losing their perfect petal shape to lie formlessly at the head of the wiry stems like splashes of dried, coagulated blood.

For John Ruskin, *Papaver rhoeas* was the flower of flowers, open as sunshine and free of 'interior secrecies' or 'coarsenesses'. There was nothing 'common' about a flower of 'silk and flame: a scarlet cup, perfect-edged all round, seen among the wild grass far away, like a burning coal fallen from Heaven's altars'. This stainless perfection made it quite a challenge for draughtsmen, but the pure red of the poppy and its sudden ubiquity in the summer months gave it the perfect entrée into spontaneous Impressionist paintings. Van Gogh painted them centre stage and in the wings: tumbling from vases, standing behind butterflies or spotting green fields with red. In the hands of his Austrian contemporary Gustav Klimt, these scarlet discs, with odd pinpoints of black, became abstract expressions of pure colour in the midst of green, white, blue and purple shapes and patterns. The American modernist Georgia O'Keeffe, on the other hand, found flowers most exciting when big and bright and seen in close up: her dazzling paintings zoomed in on the parts of plants conventionally depicted at a rather more discreet distance. O'Keeffe's *Red Poppies* captures all the energy of the roaring twenties in sinuous layers of chrome yellow and cadmium red billowing around deep, dark textured mounds. (Oddly enough, though poppies were suddenly stars of the studio, artists had been relying on them for many decades, because poppy oil, distilled from poppy seeds, was a traditional drying agent for paints.)

O'Keeffe's flamboyant painting features not the delicate, wild field flowers of Europe, but gigantic Oriental poppies (*Papaver orientale*), scarfed in vibrant red with strikingly dark spotted centres. Almost every continent has its own poppies and colours. Californian poppies (*Eschscholzia californica*) electrify summer beds with an explosion of sunshine yellow, tangerine and gold: in the Antelope Valley in California they turn the world orange, for a few weeks of the year at least. So well suited to the Sunshine State are these cheerful, laid-back, hairy flowers that it may come as a surprise to discover that their official botanical name derives from an Estonian surgeon, Johann Friedrich Eschscholz, who observed the golden fields of the Californian coastline during a Russian voyage of exploration around the Pacific Rim in 1815. In the harsher conditions of the frozen north, poppies are thinner on the ground, but lemon, white and orange Arctic poppies (*Papaver croceum*) can still be spotted blooming on bare rocks and treeless scree. Alpine poppies (*Papaver alpinum*) are slimline versions, similar in colour but with much smaller cups. The much larger, sun-loving, drought-resistant opium poppies from the Middle East (*Papaver somniferum*) often look pale in comparison and drained of energy, but they can pop out peony-like, in frilly pinks, deep purple, translucent mauve or celestial blue. There is even a variety whose white central markings and four crimson petals have inspired the name 'Danish Flag'.

With so many different colours and patterns of poppy available to modern gardeners, a well-organised bed can begin to resemble the parade of waving flags at a United Nations congress. Californian poppies, of course, are rather less formal by nature and habit, and best suited to flopping about in full sun. When my old neighbour, the retired midwife, had to move into more manageable accommodation for her later years, she chose her bungalow because of the clump of huge, red Oriental poppies with deep, dark centres beside the front door. As soon as she arrived, she scattered the seed-heads all over her new garden, and within a year or so it was as if a volcano had erupted, pouring out flames, dark rocks and molten lava over one small corner of the housing estate.

Masses of lighter lilac, purple and white opium poppies are grown in the fields around Prague, because the Czech Republic is the world's leading producer of blue poppy seeds and supplies almost a third of the global demand. The bakeries of central Europe fold in or sprinkle poppy seeds on many breads and cakes: poppy seed pastry rolls are served at Hungarian Christmases and Polish wedding breakfasts. The contrasting stripes of *Mohnkuchen*, a cake made in Austria and Germany, depend on the textured layer of speckled seeds, mixed with milk, honey and semolina. Poppy seeds are a standard ingredient of Indian cookery, too, fried with potato, stuffed into chapattis or stirred into chicken Korma. The seeds of what is now known (rather coyly) in the United States as the 'breadseed poppy' add flavour, moisture and a satisfying crunch to muffins, pancakes and lemon seed cake, as well as supplying an alternative oil for cooking and salad dressings. What seems a staple ingredient in the kitchens of some countries, however, causes deep suspicion in others. There have been reports of hapless airline passengers being arrested for failing opiate tests, when their only crime was failing to notice a few poppy seeds on their clothes; a Swiss traveller was sentenced to four years in prison in the United Arab Emirates after a very unlucky bread roll at Heathrow. The seeds of the opium poppy are also banned in Saudi Arabia, Singapore and Taiwan.

It is not the seeds that matter to the pharmaceutical industry, but the milky white juice in the bulging pods that are left after the petals drop and the seeds begin to

Seventeenth-century woodcut depicting the opium poppy offering up its valuable juices.

form. As the source of the powerful alkaloids, morphine and codeine, opium poppies have always been in high demand. The flowers were already being cultivated in ancient Mesopotamia in the fourth millennium BCE. Their narcotic properties were widely recognised in the ancient world: Roman farmers were advised not to plant these sleep-inducing plants until after the autumn equinox, when the hardest work was over. The Opium poppy *Papaver somniferum* takes its botanical name from its soporific effect, its popular name from ancient Greek ὀπός – or 'vegetable juice' – and the name of its valuable derivative, morphine, from Morpheus, the ancient god of dreams. These were the flowers that bloomed abundantly in classical mythology by the ancient cave of sleep, the home of Hypnos, where the quiet waters of Lethe flowed, inviting slumber, inducing forgetfulness. No sound of voices, no rustle of activity, no barking dogs or birdsong ever disturbed the starless darkness while the heavy scent of poppies hung suspended in the stillness. From this dark cave came Morpheus, Phobetor and Phantasos, the gods of dreams, to wait at bedsides and fill the lightless night with the shapes and colours of secret fears and desires. So powerful were the flowers of Hypnos that those who fell too deeply under their spell might never wake again.

When treated with care, the dried juice of opium poppies could offer relief from intense pain and insomnia. In the last year of his life, Dr Johnson was able to soothe some of his more troubling symptoms by 'taking the poppy' before bed and securing an undisturbed night. In eighteenth-century Britain, opium imported from India and the Middle East was widely available in medicinal tinctures, mixed with alcohol to make laudanum or 'Kendal Black Drop'. It was recommended not only for analgesic purposes, but also for treating diarrhoea, respiratory problems and coughs. When what would prove a very long war with France broke out in 1793, the perennial demand for home-grown pain relief became an urgent national need. The first successful British opium poppy field was planted the following year. At an extensive plantation in Enfield that sprang up soon afterwards, the entrepreneurial Thomas Jones decided that the best method of extraction was to set children to work, piercing the opium pods with lancets and collecting the milky juice as it dripped out. Such efficient production had the multiple benefits of reducing labour costs and dependency on expensive imports, maintaining medical supplies despite the wartime naval blockades and making Mr Jones very rich. The welfare of the child labourers seems not to have crossed his mind. Fortunately, opium never became a full-scale cottage industry in Britain, partly because of the crop's aversion to wet and windy weather.

Supplies from India and the Middle East continued to pour into apothecaries' shops and subsequently into opium cellars throughout the nineteenth century. Samuel Taylor Coleridge, prone to aches and pains, became addicted to the medicine he was given to relieve an attack of dysentery – and then suffered even more. His opium-fuelled dreams fed both the famous fragmentary vision of 'Kubla Khan' and the fiendish night terrors described in 'The Pains of Sleep'. The medicinal poppy always had two faces – one promising a blessed release from pain, the other, dangerous and deliciously addictive. In the original version of Dante Gabriel Rossetti's famous tribute to his late wife Elizabeth Siddall, the dreamlike painting of *Beata Beatrix* now in the Tate, a white poppy lies, asp-like, along her arm, perhaps

*Paris in the 1890s: an advertisement for Dr Zed's patent remedy for
infant insomnia and bronchitis, featuring opium poppies.*

to suggest the painless oblivion into which she has slipped,
perhaps to recall the laudanum that killed her.

Coleridge's young admirer, Thomas De Quincey, bought
a small dose for relieving his unbearable toothache, only
to find himself hurled into an 'abyss of divine enjoyment'.
In *Confessions of an English Opium Eater*, he devoted ecstatic
pages to his poppy-dependent pleasures, before conceding
to the profound inertia, infinite agitation and terrifying
psychotic experiences that constituted 'The Pains of
Opium'. The intense drug-induced mental disturbance
seemed to promise and deny imaginative expression. As a
fully trained doctor as well as a poet, John Keats understood
both the imaginative and sedative effects of the opium

poppy. In *The Eve of St Agnes*, Madeline undresses in her solitary chamber and, quite unaware of the hidden intruder in her bedroom closet, drifts from consciousness as 'the poppied warmth of sleep oppressed / Her soothed limbs'. His ode 'To Autumn' conveys the profound quietness of a season depicted 'on a half-reaped furrow sound asleep, / Drowsed with the fume of poppies'. Although these seem to be the common poppies of the English countryside, their 'fume' smacks of something more sinister: the reaper's work is not yet complete and autumn's induced slumbers are the prelude to oblivion. The attraction and repulsion exerted by poppies is expressed in full technicolour in *The Wizard of Oz*, when Dorothy, within sight of the Emerald City, suddenly pauses to yawn and stretch her arms, before falling, utterly overcome with sleep, into the enchanted field of deep red petals.

Morphine remains one of the most powerful painkillers in modern medicine, but because of the dangers of addiction, it is prescribed only under careful supervision. Diamorphine, the strongest of all analgesics, is used even more sparingly, while the opium poppy's most dangerous and addictive by-product, heroin, ruins lives around the world. Irving Welsh's novel, *Trainspotting*, which reached a wide audience through Danny Boyle's famous film adaptation, offers an uncompromising portrait of the drug's impact on young people in contemporary Britain, in what is at once a highly imaginative and yet horrifying series of interrelated stories. News organisations regularly present reports on contemporary drug-related tragedies from around the world. In 2017, Mexico witnessed over 25,000 murders as a result of rival drugs cartels, leading the former president, Vicente Fox to call for poppy growing to be legalised;

in the United States, the prime market for Mexican heroin, the emphasis is on prohibition rather than legalisation.

Growing opium poppies is a very lucrative business. In the remote mountainous areas of Colombia and Mexico, farmers have depended for decades on the high returns from this internationally prized crop. Opium poppies were at the centre of the recent conflict in Afghanistan, where two-thirds of the world's illegal poppies are grown. British soldiers in Helmand province initially destroyed, and then stopped destroying, the great fields of pink, white and pale purple flowers as the difficulties surrounding the immensely valuable crop became more apparent. The Afghan government were under varying degrees of pressure to control the illegal trade in heroin, but the struggle for survival in such volatile circumstances meant that the complexities of developing an adequate policy were not easily resolved. Prevention of the trade in illegal drugs may help to starve the extremists of funding, but it also exacerbates the broader suffering of a war-torn region, and a combination of poverty and resentment can be a potent agent of terrorist recruitment. The international requirement for morphine also makes legal poppy growing a potential safety line for the post-war state. The Afghan opium dispute was the latest episode in a long, unhappy history of international conflict over this compelling plant. During the nineteenth century Britain was engaged in major trade wars with China, as Indian opium, variously deployed and deplored, was the key cause of contention in what became known as the 'Opium Wars'.

With such a troubling history and international profile, it is quite disconcerting to see the opium poppy featuring in seed catalogues and garden centres along with fuchsias, delphinium and phlox. The RHS website, for example, offers strangely conventional advice on *Papaver somniferum*: 'Easy to grow in any well-drained soil in full sun or light shade . . . No pruning required but deadhead if seed is not required . . . Generally trouble free but may be attacked by aphids'. Gardens are traditional havens from the troubles of the world, but this is a remarkable diffusion of one of the world's most incendiary plants.

If modern British gardens are often riotous with bright varieties of intercontinental poppies, twenty-first-century arable farming is generally less open to colour-mixing than was once the case. There would be little chance of Madame Monet and her little boy moving through a modern field with quite the same style as in *Les Coquelicots*. Not only have fashions turned away from hats and long dresses, shawls and parasols, but the famous image also belongs to a time before farmers had succeeded in eradicating weeds from their crops. In the days of Monet, Ruskin and Van Gogh, common poppies would often transform green or golden fields into a sea of scarlet. Such natural anarchy is no longer permissible: in a well-drilled field of modern barley or wheat, identical heads stand beard to beard on matching stalks, except where tramlines have been left to allow huge agricultural machinery to fertilise and harvest the grain with maximum efficiency. The artist's inspiration is the farmer's irritant. However, wild flowers are essential to maintaining a rich ecological diversity, and the poppy is no exception. The rare poppy mason bee, *Osmia papaveris*, currently in decline across western Europe, needs poppy petals for lining its nests: it will peep out of a carefully excavated burrow, as if wrapped in a crimson robe.

Luckily, common poppies are mischievous self-seeders, so they still settle around edges and hedges, creating a colourful, speckled frame for the spotless field of gold. They often spread into hay meadows, too, from grass verges now left unmown by cash-strapped councils. Austerity economics sometimes offers unlikely opportunities for natural abundance. If these uncut roadside stretches do little to aid the visibility of drivers, they are helping to revive depleted populations of butterflies and bees. *Papaver rhoeas* is one of the most congenial plants for pollinating insects, especially as it thrives on poor soil and disturbed ground. On a demolition site where houses stand, their facades ripped away exposing far corners of wallpapered rooms, with the remnants of domestic life left open to the rain and wind and the path that once led to a front-door step now lost in a muddy quagmire, poppies are among the first plants to spring up. Construction projects, derelict farmyards, unkempt car parks or major roadworks are sites where scarlet poppies scatter and thrive.

No ground was more disturbed than the fields of Flanders in the First World War. Poppies grew where people died. The young men who were called up to serve their country were struck by the red flowers growing in what were once cornfields and orchards. In his poem on 'Break of Day in the Trenches', Isaac Rosenberg described the rat that hopped over his hand, when he reached out to pick a poppy from the parapet to stick behind his ear. This 'cosmopolitan' rat could run between the Allied and German trenches, impervious to the national differences that turned ordinary fields into hellish wasteland. The German novelist, Erich Maria Remarque, dispatched to Flanders from the opposite side of the line in 1917, would let the same rats and poppies into the novel he published a decade after the conflict, *All Quiet on the Western Front*. Isaac Rosenberg's poem ends without any attempt to make sense of what is going on around, and yet he finds a language through flowers:

> *Poppies whose roots are in man's veins*
> *Drop, and are ever dropping;*
> *But mine in my ear is safe —*
> *Just a little white with the dust.*

Rosenberg was twenty-eight when he died at the Battle of Arras on 1 April 1918. The war finally came to an end seven and a half months later.

The First World War intensified the meaning of this flower. After the war, which took the lives of millions of soldiers on all sides of the conflict, the poppy became the very image of unaccountably brief lives, of young men dying far from home. In his moving and illuminating book about nature and the Great War, *Where Poppies Blow*, John Lewis-Stempel has suggested that because numerous wild flowers, including cornflowers and daisies, flourished among the trenches, there was 'no absolute inevitability in the rise to pre-eminence of the poppy as a flower of remembrance'. The field poppy, however, did have special significance for British soldiers, not necessarily as a symbol of protest against the war or their leaders, but as an image of their own lives: 'Short, brave, brilliant'. What is even more apparent in retrospect, however, is the lack of agency suggested by the rapid bloom and disappearance of this vibrant annual.

The poppy had been linked with lives cut short for centuries before the trench warfare of the Western Front. In fact, this brief flowerer has been associated with young men sent off to sleep too soon for as long as poets have been reflecting on wars. In *The Iliad*, Homer described the death of the young, beautiful Trojan Prince Gorgythion falling 'as full blown Poppies overcharg'd with Rain' sink to the ground. Poets ever since have been quick to follow, understanding and deepening the universal language of the red flowers. Homer's poignant simile echoed in Virgil's lament for Euryalus, the young warrior whose perfect, blood-covered limbs and drooping head were reminiscent of poppies brought low by a heavy shower. Whether or not you are familiar with the classics, the sight of a clump of poppies blasted by sudden rain, perfect petals ragged and blanching, might well prompt thoughts of a battalion of scarlet coats lying on a battlefield. Sometimes the cups seem to fill with tears before they fall apart and fade away. The poppy's perennial associations with sleep and pain relief amplify its meanings for the bereaved. Red petals may come to mind in the face of a bloody, premature death, but the flower also carries more soothing thoughts of painlessness and undisturbed sleep. The soft folds of poppy petals are like layers of tissue, open and ready to wrap weary – or injured – bodies in sleep. In a moving elegy on his brother Christopher, who was run over at the age of four, Seamus Heaney drew on this potent symbol of a life stopped far too soon, recalling the sight of the little boy in his cotlike coffin, with a 'poppy-bruise' on his temple.

The poppy as a symbol of fallen youth has been recognised everywhere since December 1915, when the poem beginning 'In Flanders fields the poppies blow / Between the crosses, row on row' was published in the Christmas edition of *Punch*. It was written by the Canadian army medical officer, John McCrae, in response to the death of a fellow soldier, killed at the second Battle of Ypres. It rapidly became the most popular poem of the war and still carries a powerful resonance. McCrae's lines were read by

Dame Helen Mirren in July 2017, before a shower of poppy petals rained down from the Menin Gate over all those who had gathered to remember Passchendaele. Commemorative poppies surround the candle-lit grave of the Unknown Soldier in Westminster Abbey throughout the year and emblazon lapels and war memorials every November. They were originally promoted by the American humanitarian campaigner, Moina Michael, who was deeply affected by the vast cemeteries of northern France and pledged to fulfil McCrae's promise of keeping faith with the dead. After the end of the war, people began buying poppies in November and decorated their homes with them at Christmas. In 1921, the wreath laid at the Cenotaph by George V on Armistice Day included Flanders poppies – and has done so ever since. In the years following the First World War, more baby girls were named Poppy than during the conflict, in remembrance of the lost sons and brothers who never became fathers, of the husbands and fathers who never saw their children again.

People still wear poppies in November, out of season, in bleak, damp weather, and lay wreaths to recall the red rain and the mud. Perhaps some ancient pastoral impulse is woven into these formal rituals of remembrance? Poppies bloom and are blown in no time, but new flowers reappear the following year, as fresh as those from which they seeded. As the years pass, peaceful sleep has come to seem more possible for the soldiers who died between 1914 and 1918. But something in these flat, red reminders, simple as a child's cut-out, still refuses consolation and silently insists that even when life seems to stretch out for ever, it is really as fragile as a wild flower.

Ghost orchids, among Britain's rarest and most unpredictable flowers, appear unexpectedly in certain woods and then disappear again.

Ghost Orchids

Ghost Orchids grow under the cover of darkness, haunting hidden woodlands, evading all eyes. These rarest of British wild flowers may appear tomorrow, or they may never appear again. Dedicated orchid spotters patiently wait for the coming of the phantom flowers. Can you believe in something you've never seen and probably never will? Why not? We know they've been there before, visiting the air for a few precious days, before vanishing without trace. When botanists and conservationists gave up their vigil in 2005, ghost orchids were declared extinct. And then a few years later they were spotted once again, tiny miracles of light. They'll come again, when they're ready.

Seed Lists

Addison, Josephine, *The Illustrated Plant Lore* (London: Sidgwick & Jackson, 1985)

Andersen, Hans, *Fairy Tales*, illustrated by Arthur Rackham (London: George C. Harrap, 1932)

Anderson, E. B., Margery Fish, A. P. Balfour, Michael Wallis and Valerie Finnis, *The Oxford Book of Garden Flowers* (London: Oxford University Press, 1964)

Anon., *Emblems and Poetry of Flowers* (London, 1847)

——, *The Island of Montserrat: Its History and Development Chiefly as regards its Lime-Tree Plantations*, 3rd edn (Carlisle, 1882)

——, *The Language of Flowers: An Alphabet of Floral Emblems* (London, 1858)

——, 'The Naturalist's Diary', *Times Telescope for 1822: Or a Complete Guide to the Almanack* (London: Sherwood, Neely & Jones, 1822)

——, *The Primrose League Handbook* (London, c. 1884)

——, *The Primrose Magazine* (1887)

——, *The Primrose Picturebook* (London: Ward Lock, n.d., c. 1877)

Arnold, Matthew, *Poetical Works*, ed. C. B. Tinker and H. F. Lowry (London: Oxford University Press, 1950)

Bacon, Francis, *Essays* (London: J. M. Dent, 1906)

Barker, Cicely M., *The Book of the Flower Fairies* (London and Edinburgh: Blackie, 1927)

Bate, Jonathan, *The Song of the Earth* (London: Picador, 2000)

Bates, H. E. and Agnes Miller Parker, *Through the Woods*, 2nd edn (London: Victor Gollancz, 1969)

Bean, W. J., *Trees and Shrubs Hardy in the British Isles*, 6th edn, 3 vols (London: John Murray, 1936)

Beer, Gillian, *Darwin's Plots*, rev. edn (Cambridge: Cambridge University Press, 2009)

Belvoir Fruit Farms, Elderflower, https://www.belvoirfruit-farms.co.uk/elderflower/

Bisgrove, Richard, 'The Colour of Creation: Gertrude Jekyll and the Art of Flowers', *Journal of Experimental Botany*, 64, no. 18 (2007), pp. 5783–9

Bishop, Matt, Aaron Davis and John Grimshaw, *Snowdrops: A Monograph of Cultivated Galanthus* (Maidenhead: Griffin Press, 2001)

Blunt, Wilfrid, *The Art of Botanical Illustration* (London: Collins, 1950)

——, *The Compleat Naturalist: A Life of Linnaeus*, 3rd edn (London: Frances Lincoln, 2002)

Brickell, Christopher (ed.), *The Royal Horticultural Society A–Z Encyclopedia of Garden Plants* (London: Dorling Kindersley, 1996)

Brown, Terence, *The Life of W. B. Yeats* (Oxford: Blackwell, 1999)

Burnett, Frances Hodgson, *The Secret Garden*, illustrated by Charles Robinson (London: Heinemann, 1911)

Burns, Robert, *The Poems and Songs of Robert Burns*, ed.

James Kinsley, 3 vols (Oxford: Clarendon Press, 1968)

Campbell-Culver, Maggie, *The Origin of Plants* (London: Headline, 2001)

——, *A Passion for Trees: The Legacy of John Evelyn* (London: Eden Project Books, 2006)

Carroll, Lewis, *Alice in Wonderland* (London: Macmillan, 1865)

Carson, Ciaran, *Fishing for Amber* (London: Granta, 2000)

Chambers, Robert, *Popular Rhymes of Scotland* (Edinburgh, 1843)

Chaucer, Geoffrey, *The Complete Works of Geoffrey Chaucer*, ed. F. N. Robinson, 2nd edn (London: Oxford University Press, 1966)

Clare, John, *The Natural History Prose Writings*, ed. Margaret Grainger (Oxford: Clarendon Press, 1983)

——, *By Himself*, ed. Eric Robinson and David Powell (Manchester: Carcanet, 1996)

——, *The Poems of the Middle Period 1822–1837*, ed. Eric Robinson, David Powell and P. M. S. Dawson, 5 vols (Oxford: Clarendon Press, 1996; 1998; 2003)

Clarke, Gillian, 'Miracle on St David's Day' http://www.gillianclarke.co.uk/gc2017/miracle-on-st-davids-day/

Clifford, Sue and Angela King, *England in Particular* (London: Hodder & Stoughton, 2006)

Coleridge, S. T., *The Complete Poetical Works*, ed. E. H. Coleridge, 2 vols (Oxford: Clarendon Press, 1912)

Common Ground, https://www.commonground.org.uk

Coombes, Allen J., *The Book of Leaves*, ed. Zsolt Debreczy (London, Sydney, Cape Town, Auckland: New Holland, 2011)

Cornish, Vaughan, *Historic Thorn Trees of the British Isles* (London: Country Life, 1941)

Cowper, William, *The Poems of William Cowper*, ed. John D. Baird and Charles Ryskamp, 3 vols (Oxford: Clarendon Press, 1980–95)

Crane, Walter, *Flora's Feast* (London: Cassell, 1889)

——, *Flowers from Shakespeare's Garden* (London: Cassell, 1906)

Crawford, J. H., *Wild Flowers of Scotland* (Edinburgh: John MacQueen, 1897)

Crichton Smith, Iain, *New Collected Poems*, rev. edn (Manchester: Carcanet, 2011)

Culpeper, Nicholas, *The Complete Herbal* (London, 1653)

——, *Culpeper's Complete Herbal and English Physician* (Manchester, 1826)

Daffseek – Daffodil Database, https://daffseek.org/

Daniels, Stephen, *Humphry Repton* (New Haven, CT and London: Yale University Press, 1999)

Darwin, Charles, *On the Origin of Species*, ed. J. W. Burrow (Harmondsworth: Penguin, 1968)

——, *Autobiographies*, ed. M. Neve and S. Messenger (London: Penguin, 2002)

Darwin, Erasmus, *The Botanic Garden* (London, 1791)

David Austin Rose Nursery, https://www.davidaustinroses.co.uk/

Davidson, Alan, *The Oxford Companion to Food*, 2nd edn (Oxford: Oxford University Press, 2006)

De Almeida, Hermione, *Romantic Medicine and John Keats* (New York and Oxford: Oxford University Press, 1991)

De Navarre, Marguerite, *The Heptameron*, tr. P. A. Chilton (Harmondsworth: Penguin, 1984)

Devon Biodiversity and Action Plan, *Primrose* (Devon County Council, May 2009), http://www.devon.gov.uk/dbap-plants-primrose.pdf

Dick, S. and Helen Allingham, *The Cottage Homes of England* (London: Edward Arnold, 1909)

Dickens, Charles, *Sketches by Boz*, ed. D. Walder (London: Penguin, 1995)

Drinkwater, John, *Inheritance* (London: Ernest Benn, 1931)

Dumas, Ann and William Robinson, *Painting the Modern Garden: Monet to Matisse* (London: BNY Mellon and Royal Academy of Arts, 2016)

Eco, Umberto, *The Name of the Rose*, tr. William Weaver, Richard Dixon, rev. edn (London: Vintage, 2014)

Edlin, H. L., *Collins Guide to Tree Planting and Cultivation* (London: Collins, 1970)

Eliot, T. S., *Four Quartets* (London: Faber, 1944)

Evelyn, John, *Sylva* (London, 1664)

——, *Silva*, edited with additional notes by A. Hunter (London, 1776)

Festing, Sally, *The Story of Lavender*, 3rd edn (London: Heritage House, 2009)

Frost, Robert, *The Collected Poems of Robert Frost*, ed. Edward Connery Lathem (London: Vintage, 2001)

Galloway, Peter, *The Order of the Thistle* (London: Spink, 2009)

Gayford, Martin, *A Bigger Message: Conversations with David Hockney* (London: Thames & Hudson, 2011)

Gedge, G. C., *Sunflowers* (London: Religious Tract Society, 1884)

Gerard, John, *The Herball or Generall Historie of Plants* (London: 1597)

——, *Gerard's Herball*, ed. Thomas Johnson (London: 1633)

Gilpin, William, *Remarks on Forest Scenery*, 2 vols (London, 1791)

Goellnicht, Donald C., *The Poet-Physician: Keats and Medical*

Science (Pittsburgh, PA: University of Pittsburgh Press, 1984)

Goody, Jack, *The Culture of Flowers* (Cambridge: Cambridge University Press, 1993)

Gordon, R. and S. Eddison, *Monet the Gardener* (New York: Universe, 2002)

Grahame, Kenneth, *The Wind in the Willows*, illustrated by E. H. Shepherd (London: Frederick Warne, 1931)

Graves, Robert, *The White Goddess* (London: Faber, 1948)

Greenaway, Kate, *Language of Flowers* (London: Routledge, 1884)

Grieve, M., *A Modern Herbal*, ed. C. F. Leyel, rev. edn (London: Jonathan Cape, 1973)

Griffiths, Mark and Edward Wilson, 'Sweet Musk Roses: Botany and Lexis in Shakespeare', *Notes & Queries*, 263, n.s. 65 (2018), pp. 53–67

Grigson, Geoffrey, *The Englishman's Flora* (London: Phoenix House, 1955)

Groom, Nick, *The Seasons* (London: Atlanta, 2013)

Hadfield, Miles, *British Trees* (London: J. M. Dent, 1957)

Hales, Gordon (ed.), *Narcissus and Daffodil, The Genus Narcissus* (London and New York: Taylor & Francis, 2002)

Hall, James, *Dictionary of Subjects and Symbols in Art* (London: John Murray, 1974)

Hamilton, Geoff, *Cottage Gardens* (London: BBC Books, 1995)

Harkness Roses, https://www.roses.co.uk/

Harland, Gail, *Snowdrop* (London: Reaktion, 2016)

Harrison, Lorraine, *RHS Latin for Gardeners* (London: Mitchell Beazley, 2012)

Harvey, John, 'Gilliflower and Carnation', *Garden History*, 6, no. 1 (1978), pp. 46–57

——, *Medieval Gardens* (London: Batsford, 1981)

Heaney, Seamus, *Death of a Naturalist* (London: Faber, 1969)

——, *Opened Ground: Poems 1966–1996* (London: Faber, 2000)

Hemery, Gabriel and Sarah Simblet, *The New Sylva* (London: Bloomsbury, 2014)

Herrick, Robert, *The Complete Poems*, ed. T. Cain and R. Connolly, 2 vols (Oxford: Oxford University Press, 2013)

Hessayon, D. G., *The Flower Expert* (London: Expert, 1999)

Hockney, David, *A Bigger Picture* (London: Royal Academy of Arts, 2012)

Hopkins, G. M., *The Poems of Gerard Manley Hopkins,* ed. W. H. Gardner and N. Mackenzie, 4th edn (Oxford: Oxford University Press, 1970)

——, *The Collected Works of Gerard Manley Hopkins*, vols I and II: *Correspondence*, ed. R. K. Thornton and Catherine Phillips; vol. III, *Diaries, Journals and Notebooks*, ed. Lesley Higgins (Oxford: Oxford University Press, 2006; 2015)

Housman, A. E., *The Collected Poems*, rev. edn (London: Jonathan Cape, 1960)

Howkins, Chris, *The Elder: Mother Tree of Folklore* (Addleston: privately printed, 1996)

Hughes, Ted, *Collected Poems* (London: Faber, 2003)

Impelluso, Lucia, *Nature and Its Symbols*, tr. Stephen Sartarelli (Los Angeles: J. Paul Getty Museum, 2004)

Ingram, David, *The Gardens at Brantwood* (London: Pallas Athene and Ruskin Foundation, 2014)

Ivybridge Heritage, Primroses from Devon, http://ivybridge-heritage.org/primroses-from-devon/

Jekyll, Gertrude and Lawrence Weaver, *Gardens for Small Country Houses*, 3rd edn (London: Country Life, 1914)

Jellicoe, G., S. Patrick Goode and Michael Lancaster, *The*

Oxford Companion to Gardens (Oxford: Oxford University Press, 1986)

Johnson, David K., *The Lavender Scare* (Chicago: University of Chicago Press, 2004)

Johnson, Hugh, *Trees*, rev. edn (London: Mitchell Beazley, 2010)

Jumbalaya, Johnny, *The Essential Nettle, Dandelion, Chickweed and Thistle Cookbook* (London: J. Jumbalaya, 2003)

Keats, John, *The Poems of John Keats*, ed. Jack Stillinger (London: Heinemann, 1978)

——, *The Complete Poems*, 3rd edn (Harmondsworth: Penguin, 1988)

Kelly, Theresa M., *Clandestine Marriage* (Baltimore. MD: Johns Hopkins University Press, 2012)

Kiftsgate Court, Kiftsgate Rose, http://www.kiftsgate.co.uk

Kilvert, Francis, *Kilvert's Diary 1870–1879*, ed. William Plomer (London: Jonathan Cape, 1944)

Krikler, Dennis M., 'The Foxglove, "The Old Woman from Shropshire" and William Withering', *Journal of the American College of Cardiology*, 5, no. 5 (1985), pp. 3A–9A

Lewis-Stempel, John, *Where Poppies Blow* (London: Weidenfeld & Nicolson, 2016)

Linné, Carl von (Linnaeus), *Elements of Botany . . . being a translation of Philosophia Botanica* (London, 1775)

Lis-Balchin, Maria (ed.), *Lavender: The Genus* Lavandula (New York: Taylor & Francis, 2002)

London Medical and Physical Journal, 35 (1816)

Longley, Michael, *Collected Poems* (London: Jonathan Cape, 2006)

Loudon, J. C., *Observations on the Formation and Management of Useful and Ornamental Plantations* (Edinburgh, 1804)

——, *Arboretum et fruticetum britannicum*, 8 vols (London, 1838)

——, *In Search of English Gardens: The Travels of John Claudius Loudon and his Wife, Jane*, ed. Priscilla Boniface (London: Lennard Books, 1988)

——, Loudon, Jane, *The Ladies' Flower-Garden of Ornamental Annuals* (London: William Smith, 1840)

Mabey, Richard, *Plants with a Purpose* (London: Collins, 1977)

——, *Gilbert White* (London: Century Hutchinson, 1986)

——, *Flora Britannica* (London: Sinclair-Stevenson, 1996)

——, *Nature Cure* (London: Pimlico, 2006)

——, *Weeds* (London: Profile, 2010)

——, *The Cabaret of Plants* (London: Profile, 2015)

McCarthy, Michael, *The Moth Snowstorm* (London: John Murray, 2015)

McCracken, David, *Wordsworth and the Lake District* (Oxford: Oxford University Press, 1985)

McNeill, F. Marian, *The Silver Bough* (Edinburgh: Canongate, 1989)

McNeillie, Andrew, *Now, Then* (Manchester: Carcanet, 2002)

Mahood, Molly, *The Poet as Botanist* (Cambridge: Cambridge University Press, 2008)

Mancoff, Debra, *Sunflowers* (London and New York: Thames & Hudson, 2001)

Marder, Michael, *Plant Thinking* (New York: Columbia University Press, 2013)

——, *The Philosopher's Plant: An Intellectual Herbarium* (New York: Columbia University Press, 2014)

Martin, Martin, *A Description of the Western Islands of Scotland*, 2nd edn (London, 1716)

Martin, W. Keble and G. T. D. Fraser, *Flora of Devon* (Arbroath: Association for the Advancement of Science, Literature and Art, 1939)

Marvell, Andrew, *The Poems of Andrew Marvell*, ed. Nigel Smith (London: Routledge, 2006)

Massie, Allan, *The Thistle and the Rose* (London: John Murray, 2005)

Miles, Archie, *A Walk in the Woods* (London: Frances Lincoln, 2009)

Miller, Andrew, *Snowdrops* (London: Atlantic, 2011)

Mills, A. D., *A Dictionary of British Place Names*, rev. edn (Oxford: Oxford University Press, 2011)

Mills, Christopher, *The Botanical Treasury* (London: André Deutsch, 2016)

Milton, John, *The Complete Poems*, ed. J. Carey and A. Fowler, 2nd edn (London: Longman, 1998)

Mitford, Mary Russell, *Our Village* (London: Macmillan, 1893)

Moore, Anne Carroll, *The Art of Beatrix Potter*, rev. edn (London and New York: Frederick Warne, 1972)

Morris, Janine, *Primula Scotica Survey 2007–8* (Caithness: Scottish National Heritage Report no. 312)

Morris, William, *The Collected Works of William Morris*, 24 vols (Cambridge: Cambridge University Press, 2012)

——, *The Defence of Guenevere, and Other Poems*, William Morris Archive, http://morrisedition.lib.uiowa.edu/Poetry/Defence_of_Guenevere/defencenotes.html

Newlyn, Lucy, *William & Dorothy Wordsworth: All in Each Other* (Oxford: Oxford University Press, 2013)

Opie, Iona and Peter (eds), *The Oxford Dictionary of Nursery Rhymes* (London: Oxford University Press, 1951)

O'Reilly, Shelley, 'Identifying William Morris's "The Gilliflower of Gold"', *Victorian Poetry*, 29, no. 3 (1991), pp. 24–6

Otten, Charlotte, 'Primrose and Pink in "Lycidas"', *Notes & Queries*, n.s. 31, no. 3 (1984), pp. 317–19

Ovid, *Metamorphoses*, tr. Mary M. Innes (Harmondsworth: Penguin, 1955)

Parker, Dorothy, *Complete Poems* (London: Penguin, 1999)

Partridge, S. C., *Snowdrops: Life from the Dead* (London: S. C. Partridge, 1877)

Peacock, Molly, *The Paper Garden* (New York: Bloomsbury, 2010)

Phillips, Henry, *Flora Historica*, 3 vols (London: 1824)

Plaitakis, A. and R. C. Duvosin, 'Homer's Moly Identified as *Galanthus nivalis* L.: Physiologic Antidote to Stramonium Poisoning', *Clinical Neuropharmacology*, 6, no. 1 (March 1983), pp. 1–5

Plantlife, *Bluebells for Britain: Report on the 2003 Bluebells in Britain Survey*, https://www.plantlife.org.uk/application/files/6014/8155/5822/Bluebells_for_Britain.pdf

Potter, Beatrix, *The Tale of Jemima Puddle-duck* (London: Frederick Warne, 1908)

Potter, Jennifer, *The Rose* (London: Atlantic, 2010)

——, *Seven Flowers and How they Shaped the World* (London: Atlantic, 2013)

Preston, C. D., 'The Distribution of the Oxlip *Primula elatior* (L.) Hill in Cambridgeshire', *Nature in Cambridgeshire*, 35 (1993), pp. 29–60

Primula, About Primula, http://primula.co.uk/about-primula/

Proust, Marcel, *Remembrance of Time Past*, tr. C. Scott Moncrieff and Stephen Hudson, 12 vols (London: Chatto & Windus, 1941)

Rackham, Oliver, *Trees and Woodland in the British Landscape*, rev. edn (London: J. M. Dent, 1990)

Raven, Sarah, *Wild Flowers* (London: Bloomsbury, 2012)

Reeve, Glynis, *A Sacred Place of Elder Trees: A History of*

Tresco in the Isles of Scilly (London: Historic Occasions, 1995)

Remarque, Erich Maria, *All Quiet on the Western Front*, new edn (London: Penguin, 1996)

Robertson, Pamela, *Charles Rennie Mackintosh: The Art is the Flower* (London: Pavilion, 1995)

Robinson, Phil, *Under the Punkah* (London: Sampson Low, 1881)

Roebuck, P. and B. S. Rushton, *The Millennium Arboretum* (Coleraine: University of Ulster, 2002)

Rose Annual, The (National Rose Society of Great Britain, 1960–)

Rose, Francis, *The Wild Flower Key*, rev. Clare O'Reilly (London: Frederick Warne, 2006)

Rosenberg, Isaac, *The Poems and Plays of Isaac Rosenberg*, ed. Vivien Noakes (Oxford: Oxford University Press, 2004)

RosesUK, https://www.rosesuk.com

Royal Horticultural Society, https://www.rhs.org.uk/plants

Ruskin, John, *The Library Edition of the Works of John Ruskin*, ed. E. T. Cook and A. Wedderburn (London: George Allen, 1903–12)

Russell, Myrtle, *Lavender and Old Lace* (New York: Grosset & Dunlap, 1902)

Sacks, Peter, *The English Elegy* (Baltimore, MD: Johns Hopkins University Press, 1985)

Shelley, P. B., *Shelley's Poetry and Prose*, ed. D. Reiman and S. Powers (New York: W. W. Norton, 1977)

Sherwood, Shirley and Martyn Ryx, *Treasures of Botanical Art* (London: Kew Publishing, 2008)

Shteir, Ann B., *Cultivating Women, Cultivating Science* (Baltimore, MD: Johns Hopkins University Press, 1996)

Spry, Constance, *Flowers in House and Garden* (London: J. M. Dent, 1937)

——, *Winter and Spring Flowers* (London: J. M. Dent, 1951)

——, *A Constance Spry Anthology* (London: J. M. Dent, 1953)

——, *How to do the Flowers* (London: J. M. Dent, 1953)

——, *Party Flowers* (London: J. M. Dent, 1955)

——, *Simple Flowers* (London: J. M. Dent, 1957)

Stedman, Edmund Clarence (ed.), *A Victorian Anthology, 1837–1895* (Boston, MA: Houghton Mifflin, 1895)

Stewart, Katharine, *A Garden in the Hills* (Edinburgh: Mercat, 1995)

Strong, Roy, *The Cult of Elizabeth: Elizabethan Portraiture and Pageantry*, 3rd rev. edn (London: Pimlico, 1999)

Strutt, Jacob, *Sylva Britannica: Or Portraits of Forest Trees* (1822), enlarged edn (London, 1830)

Swarb.co.uk, Taittinger and Others v. Allbev Ltd and Another: CA 30 JUN 1993, https://swarb.co.uk/taittinger-and-others-v-allbev-ltd-and-another-ca-30-jun-1993/

Taylor, G. C., *The Modern Garden* (London: Country Life, 1936)

Tennyson, Alfred, *The Poems*, ed. Christopher Ricks, 2nd edn (London: Longman, 1969)

Thistle, The Order of the, *The Thistle Chapel* (Edinburgh: Order of the Thistle, 2009)

Thomas, Edward, *Collected Poems* (London: Faber, 1920)

Thomas, Keith, *Man and the Natural World* (London: Allen Lane, 1983)

Thornton, Robert John, *A New Family Herbal* (London: B. and R. Crosby, 1810)

——, *The Temple of Flora*, ed. Werner Dressendörfer (Cologne: Taschen, 2008)

Transnational Institute, *Poppies, Opium and Heroin Production in Colombia and Mexico*, https://www.tni.org/en/publication/poppies-opium-and-heroin-production-in-colombia-

Turner, Roger, *Capability Brown and the Eighteenth-Century English Landscape*, 2nd edn (Chichester: Phillimore, 1999)

Tusser, Thomas, *Five Hundred Points of Good Husbandry* (1573); with introduction by Geoffrey Grigson (Oxford: Oxford University Press, 1954)

United Nations, World Drugs Report for 2017, https://www.unodc.org/wdr2017/index.html

Van Gogh, Vincent, *The Letters of Vincent Van Gogh*, ed. Ronald de Leeuw, tr. Arnold Pomerans (London: Penguin, 1997)

Vickery, Roy, *A Dictionary of Plant-lore* (Oxford: Oxford University Press, 1995)

Virgil, *Eclogues, Georgics, The Aeneid*, tr. H. R. Fairclough, 2 vols, rev. edn (Cambridge, MA and London: Harvard University Press, 1923)

Walter, George (ed.), *The Penguin Book of First World War Poetry* (London: Penguin, 2004)

Warren, Piers, *British Native Trees: Their Past and Present Uses* (Norwich: Wildeye, 2006)

Waugh, Evelyn, *Brideshead Revisited*, rev. edn (London: Chapman & Hall, 1960)

Westwood, Jennifer and Jacqueline Simpson, *The Lore of the Land* (London: Penguin, 2005)

White, Gilbert, *The Natural History of Selborne*, ed. Anne Secord (Oxford: Oxford University Press, 2013)

Whitfield, B. G., 'Virgil and the Bees', *Greece and Rome*, 3, no. 2 (October 1956), pp. 99–117

Wiesenthal, Simon, *The Sunflower*, new edn (New York: Schocken, 1978)

Willis, David, *Yellow Fever* (2012), http://dafflibrary.org/wp-content/uploads/Yellow-Fever.pdf

Withering, William, *Arrangement of British Plants*, 3rd edn, 4 vols (London, 1796)

Wood Database, http://www.wood-database.com/

Woodland Trust, *Broadleaf*, 2005–18, The Big Bluebell Watch, https://www.woodlandtrust.org.uk/visiting-woods/bluebell-watch/

Wordsworth, Dorothy, *The Grasmere Journals*, ed. Pamela Woof (Oxford: Clarendon Press, 1991)

Wordsworth, William, *Guide to the Lakes*, ed. Ernest de Selincourt (London: Frances Lincoln, 2004)

——, *William Wordsworth: Twenty-First Century Oxford Authors*, ed. Stephen Gill (Oxford: Oxford University Press, 2011)

Wordsworth, William and S. T. Coleridge, *Lyrical Ballads*, ed. Fiona Stafford (Oxford: Oxford University Press, 2013)

Yeats, W. B., *The Poems*, ed. Daniel Albright, rev. edn (London: J. M. Dent, 1994)

Young, John, *Robert Burns: A Man for All Seasons* (Aberdeen: Scottish Cultural Press, 1996)

Acknowledgements

Earlier versions of some of the material which now provides sections of 'Daffodils', 'Bluebells', 'Daisies', 'Roses', 'Lavender', 'Sunflowers' and 'Poppies' were originally part of two series on 'The Meaning of Flowers' produced by Bona Broadcasting for BBC Radio 3's *The Essay*, first broadcast in September 2016 and November 2017. My thanks are due to the producer, Turan Ali, for his inspired insights and enthusiasm in making the programmes, to Emma Horrell at Bona Broadcasting and to Matthew Dodd at Radio 3, who commissioned the two series.

I have been fortunate to benefit from the expertise of the wonderful team at John Murray, including Abigail Scruby, Caroline Westmore, Hilary Hammond, Sara Marafini, Rachel Southey, Juliet Brightmore and my editor, Mark Richards. I am very grateful to all concerned, and especially to Clare Alexander at Aitken Alexander Associates for her unfailing support throughout. I would also like to record my thanks to Karen Mason at Somerville College, for her practical assistance at crucial moments.

My husband, Malcolm Sparkes, has worked very hard in the gardens we have been lucky enough to develop and nurture together, as well as creating the image of the 'Ghost Orchids' in the last chapter of this book and the flower ornaments for each of the chapter headings. Our children, Dominic and Rachael, have been crucial to our gardens and often to my own experience of gardens, great and small, of fields and riverbanks, woods and mountains and coastal walks.

We are all deeply indebted to the countless gardeners, farmers, foresters, estate managers, trusts and landowners, whose beautiful flowers, wild and cultivated, have been open to everyone to enjoy. I am especially appreciative, too, of all the writers and artists, poets and musicians, storytellers and sculptors who have helped to open people's eyes to the beauties of the natural world, as well as to the great plant-collectors, botanists, herbalists, plant scientists, horticulturalists, conservationists, curators, medical historians, cultural historians and librarians whose work has done so much to deepen my understanding of flowers.

The books, papers and websites that have been most helpful in writing this book are included in the 'Seed Lists', but it has not been possible to list all the gardens and special places to which I am indebted – though some are mentioned in particular chapters. This book is variously indebted to many friends and colleagues, and in particular Matt Larsen-Daw, Clare Greenhill, Mark Griffiths, Nick Groom, Judith Hammond, Gabriel Hemery, Anne Longley, Richard Marggraf Turley, Andrew McNeillie, Bernard O'Donoghue, Christiana Payne, the late Gwen Read, Nicholas Roe and Edward Wilson.

The seeds of this book were planted in my earliest years,

when I was lucky enough to be surrounded by people for whom flowers were all-important. My grandparents set up Bradley Nurseries in north Lincolnshire in the 1930s, at the site intended for a substantial house for my grandmother's brother, Harry. The house was never built because my great-uncle Harry, who had survived the war, in which he flew with the Royal Flying Corps, was killed in a motorcycle accident in 1924. The garden, designed by Gertrude Jekyll, was already growing and continued to thrive alongside the nursery business. I am deeply indebted to my grandparents and large extended family for the blessing of flowers and gardens in my early years. These formative experiences were often shared by my siblings, Sue, Joy and Jeremy, and owed a great deal to our remarkable parents. My mother's special relationship with flowers is touched on in the introduction to this book, which is dedicated to her.

Illustration Credits

Springs: Anti-Vietnam war rally at the Pentagon, 1967. Photo © Marc Riboud/Magnum Photos, page 9.

Snowdrops: Snowdrops from *The Temple of Flora* by Robert John Thornton, *c.*1800. The Picture Art Collection/Alamy, page 12. Woodcut from Gerard's *Herball*, 1633, page 16.

Primroses: *Bank of Primroses and Blackthorn* by William Henry Hunt. Harris Museum and Art Gallery, Preston, Lancashire/Bridgeman Images, page 24. 'Primrose Day': Disraeli statue in Westminster, 1886. Illustration for *The Graphic*/Bridgeman Images, page 34.

Daffodils: 'York in Daffodil Time' British Railways poster, 1950. Getty Images, page 40. *Echo and Narcissus* by J. W. Waterhouse, 1903. Heritage Image Partnership Ltd/Alamy, page 46.

Bluebells: *The Bluebell Fairy* illustration by Cicely Mary Barker, 1923. © The Estate of Cicely Mary Barker, 1923, 1990. Reproduced by permission of Frederick Warne & Co., page 56. *The Somerset House Conference*, 1604, artist unknown. The Print Collector/Alamy, page 59.

Daisies: From *The Ladies' Flower Garden*, 1842, illustration by Jane Loudon. Victoria and Albert Museum, London/Bridgeman Images, page 68. 'Wide Oxeyes', illustration by Walter Crane from *Flora's Feast*, 1889. Mary Evans Picture Library, page 74. 'He Loves Me He Loves Me Not', magazine illustration, 1955. © The Advertising Archives/Bridgeman Images, page 77.

Elderflowers: *Sambucus nigra* from Franz Eugen Köhler's *Medizinal-Pflanzen*, 1887, page 80. 'The Elder-Tree Mother', illustration by

Arthur Rackham, 1932, for *Fairy Tales* by Hans Christian Andersen. Granger Historical Picture Archive/Alamy, page 87.

Roses: *Rosa damascena variegata*, illustration from Pierre Joseph Redouté's *Les Roses*, 1817–24. Old Images/Alamy, page 94. Detail of Queen Elizabeth I, from the Ditchley portrait by Marcus Gheeraerts the Younger, *c*.1592. Granger Historical Picture Archive/Alamy, page 106. The Queen of Heart's gardeners, illustration by John Tenniel from Lewis Carroll's *Alice's Adventures in Wonderland*, 1865. Granger Historical Picture Archive/Alamy, page 108.

Foxgloves: Sketch of foxgloves by Beatrix Potter. Courtesy Frederick Warne & Co. & The Linder Collection, page 112. Digitalis 'Fancy Mixed' seed packet. Pictures Now/Alamy, page 120.

Lavender: Essence de Lavande, Grasse, nineteenth century. Private Collection/Bridgeman Images, page 122. Advertisement for Mitcham Lavender Water, 1890s. Amoret Tanner/Alamy, page 129.

Gillyflowers: Carnation and wallflower. Mary Evans Picture Library, page 141. Oscar Wilde wearing a green carnation, *c*.1895. Science History Images/Alamy, page 147.

Lime Flowers: Illustration from John Evelyn's *Silva (Sylva): Or, a Discourse of Forest-Trees, and the Propagation of Timber in His Majesty's Dominions . . .*, 1776, page 152. Detail of lime wood carving by Grinling Gibbons in Trinity College chapel, Oxford. Photo Malcolm Sparkes, page 157.

Thistles: 'Fat Fitie', illustration by John Ruskin. Ruskin Foundation (Ruskin Library, Lancaster University), page 160. A barefoot Viking steps on a thistle, nineteenth-century engraving. Private Collection/Bridgeman Images, page 164.

Sunflowers: *A Sunflower in a Kent Garden*, woodcut by Charles Tunnicliffe. © The Estate of Charles Tunnicliffe reproduced by permission, and with thanks to the Charles Tunnicliffe Society, page 168. *Self-Portrait with a Sunflower*, 1632, by Sir Anthony van Dyck. Heritage Image Partnership Ltd/Alamy, page 174.

Poppies: Corn poppy from Benedetto Rinio's herbal *Liber de Simplicibus*, Venice, 1419. Bridgeman Images, page 182. Early seventeenth-century woodcut of opium poppy extraction. Contraband Collection/Alamy, page 187. Advertisement for a popular French medicine, *c*.1890. Contraband Collection/Alamy, page 190.

Ghost Orchids: Illustration by Malcolm Sparkes, page 198.

Chapter vignettes by Malcolm Sparkes.

Index

NOTE: Page numbers in *italic* refer to captions to illustrations